Responsive Ethics and Participation

Responsive Ethics and Participation

Innovation and Responsibility Set
coordinated by
Bernard Reber and Robert Gianni

Volume 9

Responsive Ethics and Participation

Science, Technology and Democracy

Kalli Giannelos
Bernard Reber
Neelke Doorn

iSTE

WILEY

First published 2022 in Great Britain and the United States by ISTE Ltd and John Wiley & Sons, Inc.

ISTE Ltd
27-37 St George's Road
London SW19 4EU
UK

www.iste.co.uk

John Wiley & Sons, Inc.
111 River Street
Hoboken, NJ 07030
USA

www.wiley.com

Library of Congress Control Number: 2022939233

British Library Cataloguing-in-Publication Data
A CIP record for this book is available from the British Library
ISBN 978-1-78630-756-9

Contents

Foreword

Participatory democracy is usually promoted as a key instrument in the leveraging of the responsibility and accountability of science and technology developments regarding society. Science and technology, or "research and innovation" (R&I), covers virtually all innovation developments – of either a scientific or technological nature – that can be applied to any sector. Past studies have revealed the great diversity of participatory configurations and the varying quality of participation, leading some to consider that participation's added value in R&I might not be an intrinsic attribute. This calls into question the grounds on which ethical participation can be established, as Pellé and Reber (2016) have already shown.

What is usually covered by "participation", "engagement" or "involvement" in R&I can vary according to the context and the normative understandings, thereby generating a homogeneous representation of practices that are not always clearly identified or convergent. Participation remains a gray area in R&I and even in responsible research and innovation (RRI), where several mechanisms compete with the same promise to legitimize and offer a co-creation process or dialog between science (and technology) and society. Moreover, new technologies are ambiguous and complex, asking reconsideration of their relationships within humans (Guchet, 2021).

Taking a step backward and considering participatory practices from a governance perspective, echoing reflexive (Maesschalck, 2017) and pluralist governance (Reber, 2016), this book studies the way R&I institutions deal with participatory mechanisms and provides the tools to understand the uneven or even unstable ground forming the ethical basis for these mechanisms. Despite the great public attention and the growing institutional (and governmental) support provided to public participation in Europe, the legitimacy of these mechanisms throughout their design, implementation and follow-up is not a given. This book aims at bringing together innovation ethics with participatory practices, asking what an ethical participation could be and presenting a critical theoretical analysis of the definitions and the regulations shaping both the fields of innovation ethics and of participation in R&I. It continues the work of Sophie Pellé (2017) in business studies.

Responsibility and responsiveness imply taking a critical stance on participation, considering how participation is justified, what the goals or outcomes to be targeted are, and the nature and the scope of the underlying ethical issues. It means taking into account the preservation of the freedom and responsibilities of the actors involved (Lenoir, 2019). In addition, participatory processes relate to varying interests on behalf of the participants, thus requiring further examination of who has an interest in an R&I process, might be affected by its outcomes or might contribute to new solutions: these questions are an integral part of the legitimacy of participatory processes. At the same time, the timescale and the resources of participation have to be examined as well, as they shape the nature and degree of involvement. All of them have their part to play in the programs to develop individual and collective competencies (Loisy and Coulet, 2018).

Building on the overall blurred vision that R&I literature and practices showcase regarding participation, this book combines theoretical considerations with policy-making and empirical insights. The main orientation of this study draws from the conclusions of past publications, which have highlighted the overall lack of understanding of why, how and in view of what quality participatory practices are undertaken (Pellé and Reber, 2016). These questions are here addressed through the scope of an original methodology, empirically tested and improved, aiming at developing new taxonomies and definitions.

This book reveals the lack of common understanding around consensual terms such as “citizens”, “stakeholders”, “participation” and “ethics”. Also, an additional layer of complexity appears with the diversity of institutionalized mechanisms, some of them being formalized, others not, and with a diversified connection to ethics. It also reveals how little equipped R&I institutions can sometimes be when it comes to resources and guidance for participatory practices and for ethics (Gianni, 2016). Understanding what participation means is all the more difficult when the ethical grounds on which it can be set are not identified and when it is not clear which ethical risks or issues can be at stake.

From the perspective of R&I governance, building a framework for an ethical participation implies considering the ethical requirements of both innovation and of participation, and thereby defining a method that could allow for increased legitimacy and adaptation of participatory mechanisms. As participation requires resources (both human and financial), expertise and commitment, the need for guidance and tools to design and implement processes that are both adaptive and legitimate would be a requisite for a "responsive" participation (Reber, 2022). The analysis of the grounds on which participatory processes can be ethically developed and conducted forms the bedrock of this study.

In the governance of R&I, ethics procedures are not necessarily formalized and may not cover all processes throughout their lifetime. Similarly, when the public is invited in consultation, co-creation or any other participatory activity, the design and involvement are not necessarily part of a formalized process, nor do they necessarily abide by specific rules ensuring the quality of the initiative and its added value. From the inception to the *ex post* evaluation of the R&I outcomes, having a responsive participatory mechanism remains an ideal for which tools, understanding and capacities might be lacking.

The ideal of participatory democracy encounters profound challenges and critique over its effective connection to decision-making and regarding whether its actual impact aligns with democratic values. What if ethics and participation in R&I governance could be combined in a comprehensive and responsive system? This is the question and the guiding hypothesis of this book, which also methodically anchors the touchstone of an ethical participation. Observing how the governance of R&I struggles to harmonize the processes and the means deployed to promote public participation – citizens and stakeholders – in science and technology, it seems timely to overcome the conceptual and practical hurdles through a comprehensive approach.

The main novelty brought by this new contribution to the Innovation and Responsibility series is fourfold: first of all, it establishes an original connection between ethics and participation; second, it offers a broad overview of participatory practices, encompassing all types of activities and actors; third, it presents new categorizations and definitions within this large spectrum of participatory activities; and, fourth, it translates theory into practical guidance for ethical participation.

The approach offered by this book is general enough to be relevant for institutions related to the funding of R&I, to ethics committees and potentially to any institution or organization in the field of R&I either interested in or already implementing participatory practices. Although it is impossible to encapsulate all practices, this study gathers a variety of different approaches, methodically combined into a clarification of what an ethical participation is. Regardless of the sector or the national context, or even the objectives of a given institution, this new insight into the ethical R&I governance of participation could be of interest to anyone willing to challenge the current trends of participatory democracy.

Among the prominent features of this book, the reader will find: a state-of-the-art analysis of European regulatory frameworks on ethics and participation; a conceptual analysis of R&I ethics and of participatory mechanisms; and an attempt to merge both ethics and participation in what could be defined as ethical participation. Also, new classifications aiming at a clearer view regarding new publics invited in the arena of R&I processes are suggested, in addition to the identification of key factors worth considering for responsive participation. This book ends with a final suggestion for the design of an ethics framework, which is meant to be used as a tool for reflection and adaptation in any R&I context to properly identify who participation will be undertaken with, why and when, as well as the participatory mechanism that is relevant.

The analysis of the different publics embraced in participatory democracy leads to the distinction of different cases, such as citizens, stakeholders, organized civil society or experts, for instance, all of them being different representations of public participation. Regarding these discrepancies, the involvement of a group of participants is motivated either by their present or future implication regarding the science and technology developments at stake. This entails the identification of concerned or affected publics, with a variety of underlying justifications for their involvement. These latter can occur at different time frames, through different processes, which must be identified. Having a general overview of these processes and the tools to guide the proper "matching" between these options appears as a guarantee of the legitimacy of or ethical approach to participation. Citizen and stakeholder participation involves resources as well as a feedback and accountability which can jeopardize the legitimacy of the process if the inadequacy of the process, the time frame or the feedback loop is not tailored to fit the specific context.

The "responsive" quality of participatory practices is not an attribute that can either be part of the initial intent or part of the *ex post* justification: it implies a commitment to always address the adequacy of the participatory features, at all times (Reber, 2022). Beyond regulatory constraints or stable checklists, a responsive quality is part of a malleable set of checks and balances. As such, an ethics framework is the guarantor of a democratic approach to science and technology.

Through a state-of-the art review of the literature and of regulations, combined with an original empirical study of European research funding organizations, this book presents a novel insight into the intricacies of the governance of participatory practices. The findings of this study point out the current need for guidance in institutionalized approaches of participation, setting the ground for the shaping of ethical participatory practices.

This book aims at qualitative indicators through which the added value and legitimacy of participation can be assessed. This original viewpoint is an indirect critique of participatory inflation: the expansion of participatory democracy is not as such the precursor of a greater correlation between democracy and innovation.

To the often-encountered view of the need for "more participation", this book opposes a qualitative view, which might hinder this quantitative expansion. The main assumption is that only qualitative ethical processes can help bridge the gap between science (and technology) and society, not "more" participation (Reber, 2016). Running counter to the usual discourse on public participation, this study demonstrates why only ethical participation can be the answer in the long run. The tools and the guidance provided in this study may help reshape participatory initiatives in the governance of R&I and more broadly the way public policies refer to public participation. Demonstrating that the added value and the positive impact of participation is dependent on a qualitative and ethical approach, this book opens new avenues for qualitative investigation into the legitimacy and the long-term impact of participatory practices on individuals and society at large.

Bernard REBER
May 2022

Acknowledgments

This book stems from past work undertaken in the European project PRO-Ethics (the European Union's Horizon 2020 research and innovation program), including previous drafts and a case study that appeared in the form of project deliverables. In this respect, the authors acknowledge that this work has been possible thanks to the support provided by the European Union's funding. This book reflects the views of the authors and therefore does not necessarily reflect the views or the policies of the European Commission.

The PRO-Ethics project involved 15 European institutions, including: Sciences Po (France), TU Delft (the Netherlands), Nesta (the UK), the Danish Board of Technology (Denmark), the European Network of Research Ethics Committees (EUREC), the Centre for Social Innovation (Austria), the Eureka network and several research funding organizations from different European countries (Austria, Belgium, Czech Republic, Germany, Lithuania, Norway, Romania and Spain).

The authors of the present book would like to thank all past collaborators in the PRO-Ethics project for their insights, the real-life testing of the assumptions and the overall unique dialog that has been undertaken, both formally and informally, structuring the way toward ethical participation. For his technical help with the visualizations included in this book, the authors would like to thank Diégo Antolinos-Basso from Sciences Po's Médialab.

Also, special thanks are given to all PRO-Ethics collaborators with whom this work has been discussed, tested and adjusted: Ascensión Barajas, Reda

Cimmperman, Davide Delaiti, Madeleine Gabriel, Gerda Geyer, Radu Gheorghiu, Alex Glennie, Philipp Hövel, Xavier Hulhoven, Marie-Louise Jørgensen, Dirk Lanzerath, Zbynek Machat, Ilse Marschalek, Sabine Mayer, Niki Naska, Anila Nauni, Juliet Ollard, Ingeborg Owesen, Kristoffer Revke, Maria Schrammel, Klaus Schuch, Stefanie Schürz, Julian Stubbe, Dorothea Sturn, Lisa Tambornino, Cedric Verstraete, Erna Wenche Østrem, Martijn Wiarda, Anne Winsnes Rødland, Nyangala Zolho and many more.

External feedback has proven very useful in the orientation of the previous studies leading to this book. The authors also thank Roberta Monachello and Lisa Diependaele from the European Commission, as well as Professor Margaret Gold, for their critical reviews and positive remarks on the past work undertaken.

Last but not least, during previous research, individual interviews were conducted with renowned experts selected for their expertise in ethics and/or in participatory practices. Their views were included in a confidential report for the European Commission and have confirmed some important takes leading to the ethics framework presented in this book. We are grateful to Anne-Marie Ducroux, Vidar Enebakk, Maura Hiney, Lars Klüver, Ilse Kryspin-Exner, Mircea Leabu, Arne Manzeschke, Ana Marušić, Silvia Salinas Mulder, Tom Saunders and Loreta Tauginiene.

Introduction

Whether labeled "citizen participation", "stakeholder participation" or "public participation", participatory practices in science and technology are not easily defined as a stable process with a specific output. The participants, the format, the objectives and the outcomes differ, as well as the legitimacy of these processes and their connection to decision-making. With particularly high challenges, the field of R&I encompasses all scientific and technological fields and any sector using new technological developments – thus virtually extending to all sectors. When it comes to R&I, the participation, also termed "engagement" or "involvement", of the public (citizens and stakeholders, broadly considered) is all the more crucial as the high stakes and the potentially *affected* publics automatically justify the opening of any R&I processes to public scrutiny. The question, however, is whether R&I governance is equipped with the sufficient tools to design, implement and assess participatory processes in an ethical manner. Is participation always connected to R&I ethics? Is there any link between the ethics of R&I and participation in R&I and how could it be effectively formalized? In the governance of R&I what would be the relevance of an ethics framework for participatory practices and what would its features be? How could it embrace the geographical, sectoral, legal and cultural discrepancies in ethical considerations? Would such a framework allow for a proper guidance of participation?

Under the common labels of citizen or stakeholder participation (or engagement), multiple mechanisms challenge the way science and technology processes are usually handled with the promise of greater transparency, public acceptance and greater accountability. In line with the foundations set by the RRI framework, this book delves into the way participatory practices interfere with the co-production of solutions to global challenges and the purposeful science, upstream engagement (Wilsdon and Willis, 2004) and reflexive responsibility of scientists and innovators (Owen et al. 2012).

This theoretical investigation focuses on policy-making and therefore institutionalized processes. The interaction of innovation with society is a constant reciprocal adaptation, as scientific and technological innovation continually remakes society, which in turn accommodates, manages and redirects innovation (Guston and Sarewitz, 2002). In the fast-evolving course of R&I, participation can be leveraged as an ethical safeguard to confront new features with social and economic development. This book addresses the conditions for responsive innovation through participatory practices. Responsiveness carries the ambition of reaching R&I processes that are more adapted to the needs, values and expectations of society (Van de Poel and Sand, 2021) while including dimensions of anticipation, reflection and inclusive deliberation in policy- and decision-making (Owen et al., 2012).

Aiming at responsive ethics and participation, this study delves into the notion of responsibility and the role of ethical principles in participatory practices. Together, ethics and participation do not benefit from any straightforward connection: the latter needs to be explored and defined. Bringing together innovation ethics with participatory practices implies setting up a broad ethical perspective, preventing ethics from being reduced to formalized or standard procedures only and avoiding their occasional confusion with the legalization of ethics (soft law, ethical compliance), which also blurs the scope of their contribution. Ranging from applied ethics to meta-ethics (Reber, 2016), ethics is an extended field that cannot enter pre-designed boxes or turnkey toolkits, but as a rather open-ended approach, it helps decipher the legitimacy, the tensions and the adequacy of processes and legal compliance in regard to contextual criteria. By opening up the meanings of "participation" in R&I, this study methodically settles the boundaries of our knowledge basis: although the legitimacy of participation in R&I can be seen as self-evident, the lack of definition may lead to poorer forms of participation if the questions of purpose and the participants' roles are not clarified.

Stemming from a critical review of multi-level regulatory dimensions of participation and ethics practices, this book combines scientific and gray literature on ethical practices in R&I, unveiling general principles and methodologies of participatory practices in innovation and matching them with the basic features and concerns of contemporary ethics in the field of R&I, regulatory gaps and potential controversies. At the root of this study lies the challenge of merging the complex field of R&I ethics with the even more blurred landscape of participatory processes, as participation does not allow for a single definition. This holds true for research, innovation and new forms of democratic debates. This entanglement of meanings and practices of diverging nature could be pictured as the Tower of Babel of ethics within the specific field of participatory practices in R&I[1].

Given the very wide spectrum of the field of R&I, the present book focuses on publicly funded R&I processes from the viewpoint of governance approaches, that is, the institutions designing, conducting or implementing R&I processes. These latter can encompass projects, programs and research and development processes with any sectoral focus and any range of application. Within the governance of R&I, a specific focus is given to research funding organizations (RFOs), taking stock of their specificity in relation to the way funding can operate as a lever and to the fact that they operate with more restrictive frameworks, including standard ethical procedures.

This implies that this work does not reflect general views on innovation processes in general, but only those supported by public policies. The focus on policy-making brings into consideration the institutionalization process, while it also includes analysis that falls beyond this viewpoint. The interaction of R&I processes with society is a constant reciprocal adaptation, as scientific and technological innovation continually remakes society, which in turn accommodates, manages and redirects innovation (Guston and Sarewitz, 2002). This book explores the way participatory practices in R&I processes can be *ethically* handled, with a view to ensuring their legitimacy, the effectiveness of their action and the proper use of public funding schemes as a lever to advocate for such mechanisms in achieving better social outcomes. In the fast-evolving course of innovation – and especially technological innovation – participation can be leveraged as an ethical safeguard to confront new features with social and economic development.

---

1 Even though the scope of this study is publicly funded innovation, this analysis can also be relevant for the private sector – to the extent that it can be involved as beneficiary of or adjunct to the funding processes or to the extent that publicly funded R&I frameworks can be considered as good practices.

From the viewpoint of R&I governance, participatory democracy bears the promises of public acceptance, ensuring the legitimacy of future policies and technological developments. Although the institutional mechanisms can vary as well as the actual connection to decision-making, citizen and stakeholder participation remains a legitimate process, with few mechanisms to prevent initiatives which would not be ethically justified. With the case of RFOs, the allocation of funding schemes has to comply with criteria of fairness, equity and legitimacy. In that respect, RFOs have an advanced and arguably impartial take on the development of participatory democracy, since this latter comes as a tool rather than an external justification. This book focuses on the case of RFOs for specific views on participatory mechanisms.

The equal proportion of scientific and gray literature in this book does not quite cover some major gaps stemming from the uncertain and complex ground on which R&I evolves. The identification of varying levels of discourse also underlines this difficulty in acquiring a single view on the subject. Indeed, as some resources interchangeably use diverse terms to refer to participation, the adoption of a methodological step-by-step approach seemed indispensable. Moreover, the often vague or extremely diversified understanding of participatory practices and their use in R&I can also be problematic within the same context of actors: diverging practices and ways of considering the notion of "participation", or that of "participants", are competing. Taking stock of this difficulty, this study aims at clarifying the levels, angles and approaches, so as to better understand the standing point with regard to ethics, innovation and participation, in each discourse and practice. Despite a great number of studies undertaken at the European level in the field of ethics assessment and participatory issues in R&I[2], the extent of the remaining blind spots confirms the need to connect the findings of previous research and achievements on the matter with new questions. Accompanying gray literature on RRI and the ethics of innovation at the European level has been gathered in a selection of legal frameworks, either concerning ethics or soft law considerations on participation.

What is the responsibility of governance structures in setting R&I priorities in view of ethical concerns and participatory approaches? What is the capacity of new innovation configurations to integrate participatory

2 Especially in research funded by the European Commission through projects supported by Horizon 2020 and in particular the Science with and for Society (SwafS) program.

approaches? Should regulations be applied to new forms of participation in R&I? What should be the scope of an ethical framework for new participatory approaches in publicly funded R&I? The various underlying questions that are addressed aim to define ways to better include ethical, legal and social issues through participatory approaches in order to deliver the desired outcomes, identify the role of stakeholders as well as their involvement in R&I, and deliver insights on how R&I processes might be advanced through a proper inclusion of participation in public funding.

Further to reflections undertaken in the field of ethics of technology and also – and foremost – through the EU's RRI framework, participation is addressed here beyond its identification as a pillar of a procedural nature. While taking stock of existing reflections, regulations and policies, participation is analyzed through a methodical opening up of its dimensions and implications, also considering the very connection of ethics with participation, which is not a self-evident assumption, depending on the ways participation is dealt with. To address public concern over the impact of emerging technologies and innovations, participation can be thought beyond technology assessment procedures as they exist, in order to include a broader array of participants and allow interested or affected parties to take part in discussion and decision-making. The extension of the science–society discourse toward co-production supports a new pathway to confront the complexity and unpredictability of innovation with shared responsibility.

If ethics is considered along with participation, the reason for this interest – and the broader ambition of this book – is to respond to the great expansion of participatory practices in public policies in general, and in R&I in particular, by providing ethical guidance, which is currently lacking. This is also aligned with the current trend in RRI, which promotes a deepening of the science–society dialog through public participation. The definition of participation is far from obvious, given the very unspecific categories of "citizens" and "stakeholders"; similarly, the added value of these practices is often taken for granted and processes are applied in a one-size-fits-all manner, whereas these practices are context-dependent in their (proper) design, implementation and assessment or justification.

The first chapter of this book focuses on some theoretical grounds relating to ethics and participation, using a state-of-the art analysis of the existing scientific literature on the ethics of R&I through the scope of RRI theories and RRI policies, connecting them with theoretical considerations on participatory practices. In an attempt to define ethics and participation in the field of R&I, this theoretical section unveils the connections of participation with the ethical approach to R&I and suggests a first analysis grid based on types and the values that legitimate the conduct of participatory practices in R&I.

The second chapter explores the current European regulatory landscape on ethics and participation in R&I. This part unveils profound discrepancies between ethics regulations and the regulatory vacuum on the side of participation. These boundaries are connected to relevant features on the side of soft law that could be considered as possible pathways for future regulatory framings on ethics and participation.

Guided by the initial contribution of RRI to the promotion of public engagement as a pillar of an ethical approach to innovation, this study expands toward the analysis of all possible understandings and levels of participation. Combining theoretical resources with the analysis of existing policies, regulations and R&I contexts, the common thread in this exploration is the critical review and classification of the various overlapping dimensions. The third and last chapter examines the legitimacy and expectations from participatory practices in R&I and the ways impactful and ethical participation can be shaped in institutionalized settings – from the side of R&I governance. To do so, novel taxonomies are proposed as tools for a common framing of ethics and of participation in R&I.

Finding ways to facilitate ethically acceptable and sustainable innovation with the notion of participation involves some blind spots, such as the way to deal with minimum standards in the field of emerging technologies (e.g. artificial intelligence and bioethics), primarily concerning the boundaries of uncertainty. Considering moral pluralism and the issue of responsibility in innovation, there might be "no normative baseline on which we could judge the positive impacts and benefits of technologies"[3] – and any type of innovation, more broadly.

3 Understood here in the scope of moral pluralism. See Von Schomberg (2013).

1

Defining Ethics and Participation

The connection between science and democracy is first explored through the scope of theoretical considerations on the ethics of R&I. As a separate strand, the connection to participation is then the focus of a broad investigation of its meanings in the governance of R&I. With a view to circumscribing the links between responsibility in R&I and participatory practices, this first chapter examines ethics and participation as two separate strands with various meanings and institutional responses.

With a view to better defining, on the one hand, the ethics of R&I and, on the other hand, participatory practices in R&I, this chapter explores both dimensions in an equal manner. To do so, each strand is observed through the scope of RRI and the concept of responsibility more broadly. This diversity of ethics of R&I and participatory practices in R&I reveals a multidimensional understanding of what responsible R&I might be. While the ethics of R&I and attempts at participatory processes both overlap and are at the same time supposed to strengthen the science–society relationship, defining "ethical participation" in R&I is far from being obvious. In this perspective, trying to set the foundations of ethical participation in R&I depends closely on the identification of the ethical challenges to be found in R&I (section 1.1), on how the notion of responsibility is understood (section 1.2) and on the different understandings of participatory democracy (section 1.3). Following these distinctions, an attempt at classifying the diverse field of participation is made (section 1.4) in order to better identify the different participatory practices and their meaning.

To further define participation, and given the current theoretical gaps in the definitions, some categories are outlined as a first taxonomy. These new categories serve to provide a more granular delineation of the common understanding of public (or citizen/stakeholder) participation (or engagement), in order to clarify these blurry categories as part of a renewed definition of what is termed "participation".

1.1. Ethical challenges in research and innovation

1.1.1. *Ethics in governance of research and innovation*

Under the umbrella notion of R&I, a variety of scientific and technological fields refer to innovation strategies in a different way, each with different ethical challenges, depending on the type of funding, the end product, the beneficiaries and the maturity stage of the product or process at stake[1]. The identification of key ethical issues in R&I is inevitably a seriously flawed ambition that cannot embrace all new innovative features and the sectoral specificities that may arise. We here focus on the general ethical issues in R&I, broadly considered, and the specific needs that arise from this in terms of participatory practices.

Observing new innovation configurations (e.g. frugal innovation, grassroots innovation, inclusive innovation, open innovation and social innovation), each bears a strong yet different connection to society (the way science and technology developments connect with society). The overall goal that could be considered as a common denominator across all these forms is enhancing the democratic governance of R&I processes. Across the various emerging innovation approaches, participation is primarily taken into account in recent innovation literature, with a growing concern on deliberations and participatory approaches.

Ethical challenges can be found both in the governance of emerging technologies as well as in the implementation of new innovation modes, where they stem from the way societal challenges are dealt with, the ways of prioritizing these challenges, and the ways of addressing competitive

1 In this study, the term "R&I processes" systematically refers to any type of research and innovation process, at any stage of development and throughout the diversity of possible configurations.

interests. On the side of R&I governance, ethical features raise the challenge of reaching an efficient balance between innovation dynamics and regulatory constraints, while the specific challenges of emerging technologies raise issues regarding internationalization and the precautionary principle, which will be hereafter examined. Particularly at the European level, these ethical challenges are under scrutiny either on a flexible basis in privately financed R&I or with a high adherence in the case of R&I supported by European Union (EU) funding.

Significant public and philosophical concerns regarding the ethical, social and economic impacts of R&I are arising in many areas that could be identified as key sectors in terms of ethical/societal challenges (e.g. brain research, artificial intelligence and robotics) and emerging technologies (e.g. CRISPR/Cas9). In the face of these concerns, publicly funded R&I is confronted with the internationalization of the debate, with regulation discrepancies, ethical divides and the varying leverage of publicly funded R&I across countries. This debate raises the question of EU/non-EU relations, as the hindrance of regulations may result in the relocation of R&I activities in geographical areas not covered by applicable laws. Beneath this issue lies the already existing over-regulatory spectrum of R&I at European level: a decade ago, the European Commission Expert Group on Science and Governance stated:

> there has been a shift towards the legalization of ethics in the governance of research, which may undermine the processes of ethics in society [...] The Expert Group suggests that there has been a shift to non-binding governance or "soft law" – codes of practice, guidance, and reporting measures. (European Commission, 2010, p. 147)

This observation underlines the growing internalization of ethics in legal affairs, while, more fundamentally, it points out the blurry frontier between the ethical and the legal spheres. As ethics processes exceed the legalization of ethics, the reinforcement of this latter may hinder R&I ethics and the possibility of an ethics of participation. In the context of the internationalization of R&I, the governance of emerging technologies may find efficient tools on the side of ethics to better tackle such issues. Similarly, R&I actors can rely on ethics as they provide adapted guidance on viable pathways, encompassing both legal compliance and public acceptance.

As a first overarching commonality, all ethical considerations in R&I mainly abide by the precautionary principle, which aims at avoiding either irreversible and/or large-scale damage and ensuring security despite high scientific uncertainty. This principle is usually applied in the context of scientific controversies or in the acquisition of new knowledge. The framing of the precautionary principle is a pivotal question in R&I ethics, intervening either upon existing regulations and laws or beforehand, where regulations and laws do not exist yet (Von Schomberg, 2012a). First embedded in European policy in the 1992 Maastricht Treaty for environmental policy, it has expanded to other fields of policy (Von Schomberg, 2012a) under EU law, where it intervenes as "provisional risk management measures", before adverse effects are obvious. This principle is used in public decision-making situations, "where following an assessment of the available scientific information, there are reasonable grounds for concern for the possibility of adverse effects on the environment or human health, but scientific uncertainty persists" (Von Schomberg, 2012a, p. 147). The principle can be applied in cases: (1) of potential adverse impacts on the environment or human health with serious consequences; and (2) when governmental action should be taken regarding controversies over/lack of/insufficient scientific knowledge (Von Schomberg, 2012a). Usually, this principle applies within EU product authorization procedures, as an incentive to make safe and sustainable products (Von Schomberg, 2013). However, abuses of the principle can occur, such as disguised protectionist measures in the trade sector, which led the European Commission to set up guidelines for the precautionary principle's application (Von Schomberg, 2012a).

The precautionary principle is akin to the notion of responsibility: the key feature of its definition relates to its fundamental anticipatory nature and it "does not allow uncertainty on the scientific side of assessment to be used as an excuse when serious presumptions of significant and/or irreversible damages have been made" (Reber, 2018, p. 40). As a general rule, all cases associated with the precautionary principle imply an assessment of the state of affairs in science and technology and a review of the type of uncertainties involved, precisely because "uncertainty" can relate to several cases, depending on the context:

– *hypothetical effects and imaginary risks* requiring independent, transparent and publicly accessible scientific evaluation;

– *defined and quantified risks* requiring a risk management approach from policymakers;

– *fully or partially unreliable scientific information systems* requiring an estimation of possible adverse effects thanks to a precautionary approach;

– *impossibility of scientifically established cause–effect relationships* combined with known adverse effects requiring a precautionary approach (Von Schomberg, 2012a, pp. 151–152).

Applying the precautionary principle implies taking precautionary measures on a provisional basis. Given this time-based validity, precautionary measures are lifted when scientific knowledge has made progress to the point where former uncertainties (risk and adverse effects) turn into defined consensual levels of harm and damage (Von Schomberg, 2012a), even though it can sometimes be seen as a hindrance to innovation[2]. The risk of government reactions to emerging technologies being too early or too late can imply either failing to address their nature or missing the opportunity to intervene (Von Schomberg, 2012a). Other limitations can be found in prevailing institutional preference, when current practice is taken as the default norm among several alternatives (Von Schomberg, 2012a), which does not encompass proper assessment of future developments. Regarding normative standards, precautionary regulation implies that the standards remain open for discussion, with a view to assessing the societal acceptability of R&I outcomes.

Transposing the precautionary principle to participatory configurations, and in particular to deliberative ones, interesting interconnections can be identified, as precautionary approaches can arguably provide a "normative risk management exercise" involving "a shift in science-centered debates on the probability of risks towards a science-informed debate on uncertainties and plausible adverse effects" (Von Schomberg, 2012a, p. 156). In this respect, the application of the precautionary principle can facilitate deliberative processes (see also Reber (2016)) and further to this it could be argued, along with Von Schomberg (2012a), that, as a core ethical feature, the precautionary principle is in and of itself of a deliberative nature in that it implies deliberation on normative dimensions decided as soon as the principle connects to public policy.

2 The opposite is the "proactionary principle" promoted by transhumanists, among others, which suggests regulations should be lowered in favor of innovators. See Stilgoe (2013).

1.1.2. *Bottom-up approaches*

While there are several ways to frame innovation as a concept[3], distinctive innovation modes can be outlined as general categories, aligned with varying goals. The tendency to reduce innovations to technical innovations, prevailing since Schumpeter (Howaldt and Schwarz, 2010), has put forward prevailing patterns of innovation. With the great expansion of new innovation modes inviting parts of the civil society into the scientific and technological arenas, the so-called "bottom-up" innovation is thriving. The growing interest in such configurations lies in the renewal of the science–society relationship, with the promise of a more ethical way of conducting innovative changes.

Across the great variety of ways of defining innovation, some key characteristics can be identified in the emergence of the creation of or significant improvements to existing practices. As innovation can be applied to both an activity and its outcome, it can generally be defined as "a new or improved product or process (or combination of thereof) that differs significantly from the unit's previous products or processes and that has been made available to potential users (product) or brought into use by the unit (process)" (OECD, 2018, p. 20)[4]. Innovation creates more than a new set of techniques: it is a "future-creating activity" that brings "ultimately new social practices and even institutions that transform the ways in which human beings interact with the world around them", and by doing so "changes the world itself", either incrementally or radically (Von Schomberg, 2013, p. 119), creating added value for society (Shelley-Egan et al., 2016).

Innovation is not necessarily of a technological nature, its main characteristic being the change it brings, which can be of various kinds: from incremental to radical change, innovation can provide a different good or service (product innovation), apply new methods in the production of a good or service (process innovation) or combine a change in both a product and a process (Kaplinsky, 2014). While technological innovation creates new products, processes or features, non-technological innovations are "usually

3 The concept of innovation can be segmented into various approaches, such as the following segmentation: innovation as the development of new products; innovation as problem-solving; innovation along the steering–serendipity axis. See Khan et al. (2016).

4 According to this publication, the term "unit" refers to the actor responsible for innovations, in any sector.

social or organisational in nature" and create "new strategies, organisational forms and ideas that strengthen civil society" (Shelley-Egan et al., 2016, p. 20).

Innovation strategies can be outlined in the following four dimensions (see Francis and Bessant (2005), quoted in Bessant (2013, p. 4)):

– *product* innovation: changes in products/services;

– *process* innovation: changes in the ways in which products/services are created;

– *position* innovation: changes in the context in which the products/services are introduced;

– *paradigm* innovation: changes in the underlying mental models behind the activity.

Although the process of innovation varies depending on its nature and the actors, some main steps in producing an innovation can be identified: basic and applied research, development, scale-up or engineering, production and commercialization and dissemination and use (Dahlman and Kuznetsov, 2014). From the viewpoint of the value chain, innovation can be identified through four different dimensions: product innovation, process innovation, functional innovation and chain innovation (Kaplinsky, 2014, p. 52). This distinction allows for better identification of the involvement of participatory practices in the value chain.

In this variety of types and strategies of innovation, a great diversity of actors can intervene as main actors of the process chain: the government, state-owned enterprises (SOEs), private firms, individuals, grassroots innovators, universities and non-governmental organizations (NGOs) (Dahlman and Kuznetsov, 2014). Also, the role of public funding across the innovation value chain varies noticeably, in that across the innovation process the various actors do not have the same weight in innovation efforts – governments and private companies being the driving forces mostly (Dahlman and Kuznetsov, 2014). However, private companies often act as the main actors throughout all stages of the chain (from research to dissemination and use), the main role of government being mostly funding and performing basic research and development. Mostly represented by governments, the public sector is also well-represented at certain levels, as in funding research for general or applied knowledge, often with goals pertaining to military purposes or public health, and to some extent for industry competitiveness (Dahlman and Kuznetsov, 2014).

While the connection to new innovation configurations is variable, a sectoral view seems to weigh in favor of public scrutiny and participatory expectations in the specific case of areas of emergent science and technology (e.g. nanotechnology and synthetic biology), with acute public debate on the ways to control their development accompanied by reflection on the place of public participation both in setting research agendas and modulating research trajectories toward socially desirable ends (Fisher et al., 2006, quoted in Owen et al. (2012)). This shifts the debate from the contents toward the modes of innovation: in this sense, "the challenge is not whether or not to innovate, but *how*?" (Bessant, 2013, p. 1).

Among the great variety of innovation configurations, some of them bear a relevant connection with participation, as they embrace significant science–society interactions and a connection with participatory practices, at the process or outcome level:

– *Frugal innovation* consists of doing "more with less for more people" and encompasses many different strands and criteria. It could be considered an overarching notion rather than a sub-area of "bottom-up" innovation (Kroll et al., 2016, pp. 5–6), bringing promises of economic, social and environmental benefits (Granqvist 2016). It mainly originated in the context of emerging markets (Weyrauch and Herstatt, 2016) and, as a broad concept, it embraces various strategies, motivations and outcomes that share some common features in the product dimension, the process dimension and the context dimension (Kroll et al., 2016).

– *Grassroots innovation*: a network of activists and organizations generating novel bottom-up solutions for sustainable development and consumption, bringing solutions that respond to the local situation and the interests and values of the communities involved (Seyfang and Smith, 2007, quoted in Hossain (2016)). The grassroots innovation approach relies on needs-based user experimentation and often leads to incremental innovations (OECD, 2015). Quite similar to what has also been labeled as "crowd-based innovation" or "community-based innovation"[5], grassroots innovation is

5 Crowd-based innovation is an umbrella term that emphasizes the promise of empowering "the crowd" and which could be considered as similar to grassroots innovation in that sense. See Cuppen et al. (2019).

often devoted to energy and sustainability, although it is a broad category, encompassing other forms[6]. This bottom-up approach to innovation plays a pivotal role in sustainable development and is generated by civil society instead of government or business, as it operates without state or commercial interests. Regarding participatory processes, grassroots innovation offers interesting cases of the broadening of participation, creating linkages across sectors and across spaces, as it mainly emerges from citizens' innovations, unorganized lay people or local entrepreneurs, among others[7].

– *Inclusive innovation* focuses on the actors who benefit from, participate in and decide on the outcomes of innovation.

– *Open innovation* is based on the free flow of information and ideas across departments and organizations and is therefore a process of harnessing the distributed and collective intelligence of crowds, relying on collaboration, sharing, self-organization, decentralization, transparency in processes and a plurality of participants (Murray et al., 2010). The benefits of open innovation are better adaptation to dynamic market needs, shared resources and risks among partners and higher commercial returns. In short, open innovation can be defined as the "use of purposive inflows and outflows of knowledge to accelerate internal innovation" (Chesbrough, 2006, quoted in European Commission Directorate-General for Research and Innovation, 2016), involving all kinds of actors in the innovation process, "from researchers to entrepreneurs, to users, to governments and civil society" (European Commission, 2016, p. 86). Open innovation is "now envisaged as a quadruple helix which brings civil society into the triple helix of government, industry and universities" (Owen and Pansera, 2019, p. 39); however, the way R&I intersects with the concept of a quadruple helix and "its desire to embed civil society as a fourth strand remains little explored" (Owen and Pansera, 2019, p. 43).

– *Social innovation*: across its various forms[8], social innovation can generally be defined as "novel or more effective practices that prove capable

6 Frugal innovation strategies, for instance, may refer to a grassroots approach.

7 Also important during the scale-up phase, according to transition theory, where the engagement of citizens is essential for transition (Hossain, 2016, pp. 973–981).

8 Social innovation can refer to different types of innovation, such as social entrepreneurship (the process of creating and growing a venture, either for-profit or non-profit, where the motivation of the entrepreneur is to address social challenges); social intrapreneurship (the process of addressing social challenges from inside established organisations); and social extrapreneurship (the process of inter-organizational action that facilitates alternative

to tackle societal issues and are adopted and successfully utilised by individuals, groups and organisations concerned"[9]: these innovative practices are therefore "motivated by the goal of meeting a social need" (Mulgan et al., 2007b, p. 8). Based on social desirability, social innovation relies on participation as a process of mobilization (Howaldt et al., 2016). As such, it can be generally defined as "a process of collective creation in which the members of a certain collective unit learn, invent and lay out new rules for the social game of collaboration and of conflict or, in a word, a new social practice" (Crozier and Friedberg, 1993, p. 19, quoted in Howaldt et al., 2016, p. 27).

Each of these different approaches to innovation features some relevant science–society interactions. The connection to the ethics of participation can be considered indirectly, specifically in the quality and objectives of these interactions.

1.2. Facing responsibility

1.2.1. *The notion of responsibility in innovation*

Ethics in R&I can be broadly defined as a common platform for deliberation and discussion of values in society, based on normative perceptions, influenced by cultural norms and aiming at informing policy-making (Brom et al., 2015). Any ethical analysis in R&I encompasses specific aspects: ethical questions or the moral dilemma at stake; ethical questions per se (conformity with ethical standards); ethical issues; and ethical dilemmas when moral principles are in conflict: the evaluation of right or wrong is based on ethical/moral values (ideals), or principles and norms that define standards – identified as general ethical principles (Shelley-Egan et al., 2016). Among *ethical principles* are principles concerning individual rights and conditions deserving respect; principles concerning benefits and harms; fairness principles; and virtues (good human character traits) (Shelley-Egan et al., 2016).

---

combinations of ideas, people, places and resources to address social challenges). This typology and the related definitions are developed by Tracey and Stott (2016, p. 53).

9 Definition proposed by the Centre for Social Innovation (ZSI) based in Vienna. See: https://www.zsi.at/en/about_zsi/profile##Definition.

In the literature, key qualities of ethics in the context of R&I are:

– openness toward stakeholders and the public[10];

– public participation[11] (including information and consultation with public deliberation);

– transparency and accountability in processes;

– thematic openness in terms of which questions can be raised;

– systematic argumentation in terms of a priority of arguing over (political) bargaining (this also includes scholarly integrity).

Considering the complexity and unpredictability of innovation environments, the concept of responsibility can be considered a key dimension in innovation (Bessant, 2013), in that it ensures innovations bring benefits only. Responsible innovation (RI) stems from radical uncertainty about the future (Grinbaum and Groves, 2013): on the policy-making level, responsibility can be considered here as the frontier of existing R&I ethics with a pluralistic, evolving field of reflection. Given the future-oriented nature of innovation and its transformative power, the notion of responsibility applied to innovation refers to the future it creates, although this carries conceptual and practical difficulties (Grinbaum and Groves, 2013). This foresight issue is a problem of knowledge linked to the limits of predictability between labs and real-world implementation, the unpredictability of interference effects (Grinbaum and Groves, 2013), while it is also, ultimately, a moral problem of capacity related to human finitude:

> science is often badly suited to understanding the consequences of the actions it enables us to perform, particularly when they introduce novel entities (like nuclear reactors, genetically-modified organisms (GMOs), and so on) to the world, because important aspects of how these function in the world may not be covered by the state of the art. (Grinbaum and Groves, 2013, p. 124)

Responsibility bears an ethical stance that can refer to either the legal, moral or social spheres. Ethics and morals are often considered as interchangeable, and in either case different levels can be identified, leading

10 This could also be perceived as a participation criterion.

11 "Participation" could act as an overarching notion embracing all these ethical dimensions.

to the distinction between applied ethics, normative ethics (moral theories) and meta-ethics, despite their permeability (Pellé and Reber, 2016). Also – in addition to the pluralism of ethical levels – the pluralism of moral theories entails that although deontology prevails in our current context, several rational paths can be followed through an ethical pluralism applied to ethics assessment (in the context of justification):

– types of entities assessed from a normative ethics perspective;

– normative factors;

– foundations (foundational normative theories).

Conflicting factors or hybrid forms of reasoning pave the way toward the surpassing of regulations (as in ethics review/assessment) and deontologist ethics toward an inevitably broader scope, pluralistic, implying an enhancement of reflexivity and responsibility[12]. This latter is notably polysemic, and its historical roots are to be found in imputability and accountability (Grinbaum and Groves, 2013): "responsibility" derives from the Latin term *respondere* (to respond), which can be identified in 10 different situations or understandings: cause, blameworthiness, liability, accountability, task (role), authority, capacity, obligation, responsiveness and virtue (care).

The various meanings of responsibility can be divided between negative and positive interpretations, the first category being notably focused on the individual and causal chains and imputability, whereas the second one has a prospective element, the future being the horizon which determines morally desirable goals (Pellé and Reber, 2016). Among possible meanings of responsibility, the most relevant one appears to be the individualist and consequentialist[13] concept of responsibility that prevails, implying that moral agents anticipate the consequences of their actions: "rather than duties being pre-ordained, it is up to the individual moral subject to take responsibility for deciding what s/he should do, and to prepare to be accountable later for the consequences" (Grinbaum and Groves, 2013, p. 121). In that sense, taking responsibility is close to foresight, while ignorance at the time of the action

12 See the in-depth analysis provided on the law/ethics divide and the pluralism of ethical theories in Pellé and Reber (2016, p. 31 *sqq*).

13 The ideal consequentialist moral agent in this context might be Jeremy Bentham's utilitarian subject, who considers whether each action serves the aggregate happiness of society; see Grinbaum and Groves (2013, p. 122).

does not mean that there is no moral or legal responsibility (Grinbaum and Groves, 2013). Among these interpretations of responsibility, those pertaining to emerging technologies/forms of innovation are interpretations based on a normative appreciation of the good: responsibility as moral obligation; responsibility as responsiveness; and responsibility as virtue (as "care", as an obligation to be held accountable).

Responsible innovation "starts from an understanding of innovation as a system, a web of myriad actors, rather than a pipe" (Stilgoe, 2013, p. XV). This complex web includes users, scientists, entrepreneurs, governments and others (Stilgoe, 2013) where there is an endorsement of the relevant public values during the innovation process (Taebi et al., 2014). Responsible innovation "means taking responsibility in ways that are, respectively, quasi-parental and collectively political in nature" (Grinbaum and Groves, 2013, p. 134). On the opposite side, irresponsible innovation can occur in contexts where the importance of the innovation's societal context has been miscalculated or when unresolved conflicts have taken place during the innovation process, which usually involves several actors: this can result in four types of *irresponsible* innovations (Von Schomberg, 2013):

– technology push[14];

– neglect of fundamental ethical principles;

– policy pull[15];

– lack of precautionary measures and technology foresight.

As noted above, responsibility in and of itself has a wide array of approaches. As part of EU-funded projects, ethical research identifies the ethical issues in research projects in interaction with the scientists and can be either in parallel or embedded research, the latter implying a closer cooperation with the researchers and the advantage of having a commitment from both the ethicists to the project and from other project members to the ethical research as part of the project (Van Gorp and Van der Molen, 2011). Even though moral/ethical expertise is debatable, ethicists provide

14 In cases where substantial dissent among major stakeholders does not allow responsible technological development and one company takes the lead with a technology push. See Von Schomberg (2013, p. 61).

15 When policymakers integrate new technologies/innovations beyond their technical feasibility, proper technology assessment and public scrutiny. See Von Schomberg (2013, p. 62).

experience in identifying ethical issues and usually have some formal education in ethics that provides a specialization useful to guide research: ethicists[16] have access to all relevant material and information and help broaden the discussion and reflect on the consequences of the choices during decision-making (Van Gorp and Van der Molen, 2011).

Furthermore, the moment when ethics intervenes determines the very possibility of having an impact: as ethical guidance could be ineffective once technology has been developed or introduced in society, this urges anticipation and upstream engagement[17]. Finding the right momentum is a problem known as the Collingridge Dilemma:

> at the time when we can still make changes to the technology, one lacks the information about effects which only the introduction and use of the technology in society could provide, but at the moment that the technology has been introduced in society and information about its effects and morally salient characteristics starts to become available, it is often very hard to still make changes. We should aim to have results of ethical discussions available at a moment when they can still be used to inform the design, implementation or utilization decisions. (Van den Hoven, 2014, p. 6)

While it seems clear that responsible innovation should be directed at socially desirable and socially acceptable ends, both in concept and practice its definition remains unclear (Owen et al., 2013).

1.2.2. *Ethics and technology assessment: at the root of responsible innovation*

The field of ethics has seen a significant development in the past century. From a predominantly meta-ethical enterprise in the beginning of the 20th century, the focus in the 1960s gradually shifted to more applied forms of ethics, of which medical ethics is probably the most prominent one. In these early years of applied ethics, the focus was still mainly on the application of

---

16 Ethicists can be similar to the consultants, or researchers, who are conducting research within the considered project (Van Gorp and Van der Molen, 2011, p. 41).

17 Upstream engagement is examined in the section on new trends in innovation approaches and configurations.

ethical theories to practical problems, for instance deontology and utilitarianism (Manders-Huits and Van den Hoven, 2009). Since the 1990s, the ethics of technology has emerged as one of the important branches of applied ethics. Within this field, ethical research is increasingly carried out as the so-called "ethical parallel research": the idea behind this type of research is that ethical investigations are carried out parallel to, and in close cooperation with, a specific technological research and development project. The ethicists provide skills that scientists, researchers or innovators do not have: this interaction allows the ethicists to co-shape new technological developments. Whereas pre-World War II management was aimed at increasing the pace of technology development and early technology assessment (TA) at inhibiting the pace, the third-generation TA attempts are aimed at informing the direction of technology development (Van de Poel, 2008, quoted in Doorn and Nihlén Fahlquist, 2010).

With the growing complexity of issues of participation and governance in the mid-20th century, science and technology policy evolved toward TA, which emphasized research on the social, ethical and environmental impacts of scientific and technological change (Shrader-Frechette, 1995, pp. 2484–2490, quoted in Fisher et al., 2006, p. 487) in response to the growing societal awareness of undesirable consequences (Van Lente et al., 2017) that are bound to technological offerings, thus offering guidance for funding and regulatory activities (Shrader-Frachette, 1995). TA started with statements about the future performance of technologies, thus assessing their impact on society (Van Lente et al., 2017), and over five decades has provided a set of philosophies, practices and approaches. Impact assessment reflected the belief that mapping the consequences of future technology on all relevant dimensions was feasible and also useful, as it was supported by probabilistic methods[18].

Thanks to its anticipatory nature, TA served for forecasting and research and development governance by helping decision-making about which technologies should be funded for development or how they might be regulated. TA has put experts at the forefront, as this impact assessment was exclusively expert-led, though TA assumptions have been criticized for the level of predictability (considering nonlinear and indeterminate processes of R&I) and the legitimacy of experts' knowledge in decision-making when several values are at stake (Van Lente et al., 2017). Several forms of TA

18 Risk calculations and decision theory are part of them; see Van Lente et al. (2017, p. 254).

have been developed, categorized in the literature through distinct features, which are mainly parliamentary TA, expert TA, participatory TA, constructive TA and discursive TA.

Public concern over the impact of emerging technologies has been traditionally addressed both "downstream" of technological development (e.g. by regulations and market mechanisms) as well as "upstream" (e.g. by research policy and TA) (Fisher et al., 2006). However, TA methods became progressively more participatory and constructive following the Dutch constructive TA (CTA), with a broader array of participants, seeking to influence both upstream decisions and technological design decisions (Fisher et al., 2006). The inclusion of stakeholders in TA addressed the issue of experts' exclusive judgment and decisions in policies in order to reach a more democratic innovation policy process (Van Lente et al., 2017).

Public upstream engagement came along as a new form focusing on more interactive approaches to science–technology–society relations, using dialog and other engagement practices. This method aims at creating a communication loop, from the public to policymakers, scientists and engineers (Fisher et al., 2006): an information loop between science (and policy-making) and the public, such as the public participation model promoted by the Danish Board of Technology and its consensus conferences[19]. Public engagement methods imply the participation of scientists and engineers as citizens and a distinction can be made between upstream engagement techniques aimed at policy-making processes and midstream engagement techniques aimed more explicitly at influencing the self-governance of research and development processes (Fisher et al., 2006).

A number of science and technology projects (e.g. in information technology and nanotechnology) have engaged ethical, legal and social aspects/implications (ELSA/ELSI) in their work in order to help ensure that the production of knowledge and technology develops in accordance with social concerns and values. Similar to TA and ELSI programs, upstream approaches emphasize the early consideration of sociotechnical implications (Fisher et al., 2006) in order to intervene from the outset as real-time TA, as an observation mechanism.

19 Practiced since the 1980s. See Fisher et al. (2006, p.488).

As well as TA and ELSI, methodologies for real-time TA have been developed, bringing elements lacking from TA and leading to an inherently reflexive research and development enterprise, where social science activities are fully integrated with the core science agenda (Guston and Sarewitz, 2002). This can be achieved through[20]:

– studying past examples of transformational innovations to anticipate future interactions between society and new technologies;

– mapping the resources and capabilities of the relevant innovation enterprise to identify key research and development trends, major participants and their roles and organizational structures and relations;

– eliciting and monitoring changing knowledge, perceptions and attitudes among stakeholders;

– engaging in analytical and participatory assessment of potential societal impacts for informed societal response to innovation (from scientists to the general public)[21].

In addition to the evolution from hard impacts toward soft ones (Van Lente et al., 2017), the journey from TA to CTA and then RRI[22] has led to an internal evolution of the focus and goal of ethical assessment, and its resources and legitimacy: the former TA focused on impact, and drew from experts' knowledge, while its political legitimation was anchored in science mobilization. CTA's focus on inclusion and design is rooted in public/citizen knowledge (or TA agents), and its political legitimation lies in serving democracy and improving technology for societal reasons. However, all three forms – TA, CTA and RRI – have each been criticized for their possible biases: TA for the marginalization of relevant aspects and CTA for the intricacies of stakeholders' interests, while the next feature, RRI, is not immune to naive instrumentalization (Van Lente et al., 2017, pp. 255–256).

20 These TA approaches have been identified as such in Guston and Sarewitz (2002, p. 100 *sqq*).

21 The last two approaches have a higher degree of participation.

22 RRI is examined in the next section.

1.2.3. *From ethics of technology to responsible research and innovation*

As well as TA, responsible innovation is another strand at the intersection of ethics and technology: rooted in the ethics of technology, responsible innovation refers to the intention of making innovation consistent with important public and moral values by including these values at all stages of innovation, as a collective duty of care, with regard to both the desired outcomes of innovation and also the way to make its pathways responsive in a context of uncertainty (Owen et al., 2012). RI and RRI are parallel discourses[23], as both of them "at times variously intersect with, reinforce or challenge existing de facto narratives and norms of responsibility as these relate to scientific research, development and innovation" (Owen and Pansera, 2019, p. 26)[24]. Also, both RRI and RI are interpretively flexible and politically malleable (Owen and Pansera, 2019).

As a policy discourse that emerged from the European Commission, RRI is rooted in the European Commission's Science in Society program, and shares much with the discourse of RI, which has in contrast emerged largely from academic roots (Owen and Pansera, 2019) and remains mostly an ideal, a guiding principle, unresolved in terms of its political, institutional and normative practices (Owen and Pansera, 2019). As an open question, RI "asks how we can and should meaningfully engage as a society with the futures innovation seeks to create, futures that are being created unintentionally or by design" (Owen and Pansera, 2019, p. 28).

On the other side, RRI aims at R&I processes that are more responsive and adaptive to societal grand challenges and requires broader foresight and impact assessments (Von Schomberg, 2013), without having an innovation strategy driven on its normative side by market demand[25]. As such, it can be considered as a strategy of stakeholders to better anticipate R&I outcomes

23 It could be argued that RRI offers a breakdown of responsibility in several (external) dimensions and as such does not fully cover the notion of responsibility in and of itself; see Reber (2019, pp. 50–73).

24 The authors specifically refer to, for example, those relating to academic conduct and research integrity.

25 See René von Schomberg's analysis, who states that "the normative dimension of what counts as an 'improvement' is decided by market mechanisms" and that "positive impacts of innovations under public research and innovation policy schemes are solely justified in purely economic terms" (Von Schomberg, 2013, p. 54).

aimed at the grand challenges of our time for which they share responsibility (Von Schomberg, 2013). RRI's focus on outcomes is nurtured by ethics and aims at aligning to aspirational notions, which seals its legitimacy. It aims at several normative focal points such as, notably (Von Schomberg, 2013): ethical acceptance (compliance with fundamental values) sustainability and social desirability.

The roots of the RRI approach go back to research in applied ethics and the ethics of technology more particularly. The intersection of ethics with technological innovations, applied science and engineering provides practical innovative solutions for important social and global problems that could shift public policy and decision-making (Van den Hoven et al., 2014). In the applied ethics of technology research programs, participation (by stakeholders[26]) is considered a key element: input from civil society, consumer organizations, NGOs, decision-makers and politicians, professionals and market parties would ensure the connection with real-world matters (Van den Hoven et al., 2014). By integrating all affected parties, the assessment of innovative features necessarily embraces contextual elements, regulatory frameworks, incentives and institutional arrangements (Van den Hoven et al., 2014).

With the European Union's Horizon 2020 research strategy, the concept of RRI found its most clear policy expression, one that considers ethical, legal and policy issues early in the innovation chain, aligning R&I goals of the European Commission with broader societal needs through Horizon 2020 societal challenges themes (European Commission, 2013, quoted in Khan et al., 2016, p. 80):

> Responsible Research and Innovation is an approach that anticipates and assesses potential implications and societal expectations with regard to research and innovation, with the aim to foster the design of inclusive and sustainable research and innovation. It implies that societal actors (researchers, citizens, policy makers, business, third sector organizations, etc.) work together during the whole research and innovation

26 Representatives from the innovation and technology sector.

> process in order to better align both the process and its outcomes with the values, needs and expectations of society[27].

In its current framing in Horizon 2020, RRI includes six policy keys, one of which explicitly concerns ethics, another one public engagement[28]. The intrinsic value of ethics seems to contrast with its identification as a "key" of RRI: limitations have been reported with these keys, such as the fact they are isolated themes and do not substantively engage with innovation systems (Owen and Pansera, 2019).

As a concept, RRI has so far taken a substantial path throughout national and European policy, with a main focus on societal directions that appear to be desirable, "stressing the ethical aspect of new technologies: they do not merely produce new risks and benefits, they alter the symbolic or moral order as well", which leads to "the idea that reflection on research and innovation should incorporate normative ideals" (Van Lente et al., 2017, p. 255), as it embraces a broader set of values and focuses on morally ambiguous situations and moral controversies[29]. Early engagement of stakeholders in R&I has been progressively acknowledged in the area of policy-making. The idea behind RRI is that societal actors and innovators become mutually responsive to each other[30], thereby co-creating solutions for which they share responsibility. The notion of responsibility plays a crucial role in RRI as it establishes the link between participatory technological assessment and ethical assessment, two domains from which RRI inherits (Reber, 2018).

---

27 See https://ec.europa.eu/programmes/horizon2020/en/h2020-section/responsible-research-innovation [accessed 17 May 2020].

28 The six policy keys for RRI as set by the European Commission are: ethics, gender equality, governance, open access, public engagement and science education.

29 At the same time, RRI takes stock of soft impacts, which pertain to private sphere and individual choice matters. See Van Lente et al. (2017, pp. 256–257).

30 RRI is a transparent, interactive process by which societal actors and innovators become mutually responsive to each other with a view to the (ethical) acceptability, sustainability and societal desirability of the innovation process and its marketable products (in order to allow a proper embedding of scientific and technological advances in our society). See Von Schomberg (2012b).

The various dimensions of RRI reflect distinctive perspectives and discourses[31]:

– for policymakers: "address grand societal challenges", "increase public trust" and "build a responsible future";

– for researchers: "incorporate other views", "evaluate the impact", "anticipate, reflect, engage, act", "share the process, make it worthy" and "share your responsibility";

– for business and industry: "invite all relevant actors", "boost socially acceptable innovation", "find new business opportunities", "reinforce your customers' trust" and "add value, secure your future";

– for the education community: "responsibility is a learnt behavior", "stimulate curiosity", "contextualize science" and "empower future generations";

– for civil society organizations and RRI for citizens: both relate to "your voice and ideas are important", "co-create the future", "be informed, be critical" and "the media are key actors".

RRI is usually associated with discourses on democratic governance, responsiveness (reflection and deliberation) and responsibility[32]. The second one refers to the integration and institutionalization of established mechanisms of reflection, anticipation and inclusive deliberation in and around the processes of R&I: this answers to the need for inclusive deliberation processes of dialog, engagement and debate, inviting and listening to wider perspectives from publics and diverse stakeholders (Owen et al., 2012). The main specificity of RRI in regard to other governance concepts lies in its emphasis on deliberation about the purposes and motivations, not only products, of innovation and on responsiveness. Participation seems key in this RRI approach, as it aims at involving all stakeholders in R&I processes from an early stage on. While RRI could be considered as an improvement of TA, by giving weight to ethical deliberation, and taking stock of the ambiguous consultation of stakeholders,

31 According to the perspectives outlined in the SwafS program RRI Tools (community project for research, technological development and demonstration – rri-tools.eu), funded by the European Union's Seventh Framework Programme. In this project, six different perspectives on RRI are identified, each stressing different aspects of RRI.

32 Although this can be called into question; see Reber (2018, pp. 38–64).

RRI still has room for improvement if we consider some criticisms that have been noted.

Innovation governance goes beyond risk governance and, in the field of responsible innovation, it implies that processes and outcomes are taken into account (Von Schomberg, 2013). Among regulatory frameworks, codes of conduct have the advantage of flexibility, which can prove useful when the ground for legislative action is still uncertain, while they also help identify knowledge gaps and target societal objectives that can be useful for research funds (Von Schomberg, 2013). Responsible innovation requires responsiveness: making innovation more responsive to societal challenges entails more responsive governance choices and a better consideration of the underlying values, norms and principles behind policies (Stilgoe et al., 2013). In this perspective, responsiveness is an integral part of responsible innovation, which, according to Stilgoe et al. (2013), can be defined through four main dimensions: anticipation, reflexivity, inclusion and responsiveness. The latter implies addressing new knowledge and emerging perspectives and norms (Stilgoe et al., 2013) in view of societal challenges.

1.3. Public participation and the ideal of participatory democracy

1.3.1. *Public participation in research and innovation*

From information and consultation to more acute participatory forms, public participation in science/technology developments expands in various directions[33]. Public participation can operate in decision-making on science and technology policy, either at limited or enhanced levels, when public views are actively solicited; for instance, in strategy, program design and evaluation. Participation can refer to the involvement of stakeholders, of citizens (general public) or of specific categories of participants that are usually under-represented, as well as end-users of the targeted product/service/output of a given R&I process. These variations also cover different stages of R&I, while citizens as a wide category can refer either to individuals or to civil society actors – and citizens organized in associations

33 Although there are also participatory processes in new innovation modes (e.g. social and open innovation), which assume a much wider range of participants than traditional innovation processes; however, in these processes, the mode of participation – not exclusive to public participation – is not a formalized participatory process and not addressed as such in the reviewed literature.

too – as the potential of these latter in revitalizing participatory impulses and ideals has been acknowledged[34]. Public participation – also labeled "public engagement" – is more widely known since it has been adopted and promoted by the European Commission (Rask et al., 2018).

Public engagement might be challenging in data collection given the discrepancy that can occur between real socially desirable answers (reported public opinion) and true public opinion, in surveys for instance (Van Gorp and Van der Molen, 2011). This has been verified in several experiments, through sectoral cases, where it seems that participatory (deliberative) processes stimulate efforts to enhance desirable impacts and mitigate undesirable ones in the decision-making process set up by researchers, for example about research priorities and directions (Guston and Sarewitz, 2002).

The main formalized public participation methods are: referenda, public hearings/inquiries, public opinion/surveys, negotiated rule making, consensus conferences, citizen juries/panels, citizen/public advisory committees and focus groups. They can be grouped into five categories, regardless of the variability of the inner characteristics pertaining to each category:

– *public hearings*: a broad category which refers to several mechanisms, tending to be loosely structured as open forums, with the appearance of individual and community involvement;

– *initiative*: considered as the prototype of democratic processes, initiatives enable citizens to place issues on the ballot for voter approval;

– *public surveys*: these can complement participation through hearings or written comments by providing a more representative portrait of public opinion and by offsetting biases and seeking opinions more broadly from the uninterested but affected public;

– *negotiated rule making*: "participants generally consider the products of a negotiation to be more informed, pragmatic, and workable than products of a conventional rule making. Parties have access to information as it is needed and the opportunity to educate others and persuade them of the reasons behind their positions";

– *citizens review panels*: "participants can influence the agenda, question experts, evaluate evidence, balance competing considerations, and debate

34 On this topic; see Fung (2003, pp. 515–539).

issues, possibly with authoritative decision makers". However, this model reaches only a portion of the affected public (Fiorino, 1990, pp. 230–235).

Several other participatory methods can be identified, with different objectives, topics, categories of participants and degrees of involvement:

> citizen juries, consensus conferences, deliberative conferences, the Delphi and Charette methods, focus groups, planning committees, scenario workshops, 'visions of the future' consumer workshops, global cafés, opinion polls (with or without deliberation), questionnaires, citizen advisory committees, vote conferences, interactive TA, constructivist consumer TA, ad hoc committees relating to the rules of negotiation, interdisciplinary work groups and political role play. (Reber, 2005, pp. 812–813)[35]

The extent in the selection of main participatory methods varies across the literature, as some authors consider fewer methods or additional ones, such as planning cells, for instance Slocum (2003). Across scientific and gray literature, participatory methods considered to facilitate higher levels of involvement are: the Charrette method, the Delphi process, citizens' juries, citizens' panels, consensus conferences, deliberative polling, the round table method, scenario workshops, search conferences, study circles and methods of sustainable community development (Shelley-Egan et al., 2014). Some attempts to classify methods can be found, for instance through an analysis grid that takes stock of the objectives, the topic, the participants and the time length of the process (Slocum, 2003).

Among these different models, no single one prevails, as contextual factors determine the effectiveness of each one (Rowe and Frewer, 2000). Judging on participation criteria, these models have diverse internal characteristics: allowing direct participation of amateurs (all except negotiated rule making), exercising full decision authority (in initiatives and negotiated rule making), and allowing discussion (in negotiated rule making and citizen review panels). Also, all models do not offer the same basis of equality in terms of influence over the definition of issues: this mostly occurs in negotiated rule making, where the public's consent is a

---

35 The length of this list varies across the literature. This substantial list goes further than Slocum's 10-entry classification.

pre-requisite for the conclusion of the process (Fiorino, 1990). In addition to the limitations inherent to lower forms of citizen participation, the risk of distortion of participation also narrows the efficiency of participation in terms of actual outcomes.

Participatory processes are not obvious and public talk can be "derided as unsubstantiated words and empty rhetoric" or even occasionally perceived as an impediment to science-led progress (Irwin, 2006, p. 318). In addition to this vulnerable construct, tensions can put legitimacy at risk and undermine the ethical grounds of participatory processes:

– biases can potentially occur at any level/type of participation, with external or internal stakeholders[36];

– scientific and ethics evaluations can diverge;

– public consultation can occur without a link to decision-making;

– potential lack of dialog or knowledge sharing with lay citizens;

– potential lack of dialog on future orientations/targeted impacts.

To reach equity and fairness and avoid biases, concerns regarding ethical issues must also be addressed, taking into account their impact on society, especially for emerging technologies in fields such as medical research and nanotechnologies: new science and technology can have a considerable impact upon vulnerable groups – including the unborn and those unable to defend themselves (Sykes and Macnaghten, 2013). Empowering wider social agency in innovation through participation is all the more justified in contexts where social choices are issued from a narrow set of incumbent interests (Stirling, 2005). These latter can intervene in appraisal by framing it in a way that closes down the range of social choices: thus, a stylized participation could be opposed to an "opening up" participatory appraisal[37].

Contradicting the tendency in science study to resolve issues of governance through legal treatment (Irwin, 2006), participation shows another way in which open science–public relations can bring along

36 External stakeholders in assessment processes can represent interests as well.

37 Motivations underlying appraisal correspond to three types of imperatives that shape the role of intentionality: normative (the right thing to do), instrumental (aiming to secure particular ends) and substantive (aiming at generally better ends). See Stirling (2005, p. 264).

significant legitimacy to R&I processes. While on some topics participation is perceived through the perspective of enhancing public trust, a naive focus on engagement cannot in and of itself rebuild trust and be taken as an antidote for public skepticism over technical change and innovations (Irwin, 2006) when there is such a divide between science and society. In line with a democratic perspective, it could be argued that participation should be driven by ethical grounds targeting a greater degree of deliberation and inclusion, so as to offer a constant space for inclusive dialog.

Participation addresses general concerns of the general public on innovations' purposes and motivations, questions such as: "in whose interests is the science being developed? Are particular innovations necessary? Are there alternatives?" (Sykes and Macnaghten, 2013, p. 102). The importance of social agency in science and technology choices is of importance in policy-making, where public dialog is often claimed to increase the legitimacy of decisions. In other spheres, participation's legitimacy lies in the fact that it helps "explore the grey areas of public opinion, away from polarized discussion and media sensation, enabling more nuanced and in-depth thinking about an issue" (Sykes and Macnaghten, 2013, p. 97). The legitimacy of participatory processes depends as well on the time in which and the way they are conducted: addressing ethical dilemmas in upstream participation where science/innovation is at an early stage is, for instance, a determining factor in ensuring the ethical grounds of innovation processes.

1.3.2. *Participation in technology assessment*

Moving from loosely defined "participation" to the identification of participatory methods, one stumbles upon questions such as "what participation mechanism is most effective in enabling public participation" and "in what circumstances", while "to be able to test it, one must possess definitions of such important concepts as *participation mechanism, effective*, and *circumstances*" (Rowe and Frewer, 2005, p. 252).

Evaluation is probably the most institutionalized and long-standing field of participatory practices, notably renowned through participatory TA. Participation in TA does not cover the whole array of participatory practices in innovation but has extensive importance in ensuring the legitimacy and efficiency of funded projects. However, the distinction between *ex ante* assessment and *ex post* assessment[38] is not always specified in the scientific literature on the topic (Reber, 2005), blurring the lines of criteria and implications.

In the evolution of public engagement, the development of TA organizations, in Europe and in the United States from the 1970s onward played a key role as a stakeholder model: they provided a warning of future technological mishaps, and raised awareness on the lack of democratic input in technological governance, the answer to this issue being the presence of elected representatives at an early stage (Sykes and Macnaghten, 2013). The Danish Board of Technology led this a step further, through a more inclusive approach to assessment, not only informing Parliament but also creating a space for dialog with public debates and the related model of public deliberation, broadly understood, through consensus conferences (Sykes and Macnaghten, 2013). In addition to this development, the Netherlands fostered deliberative fora through diverse technological domains, bringing a substantial contribution to institutional innovation on public deliberation, mainly thanks to the Rathenau Institute (formerly known as the Netherlands Organization for Technology Assessment (NOTA)) (Sykes and Macnaghten, 2013). These developments were propelled by intense public debate in the media and in civil society organizations on critical issues (e.g. on genetic modification of plants and animals), which required broad societal debates (Sykes and Macnaghten, 2013). The growing concern over scientific controversies kept directing attention towards the lack of the public's confidence in science and the need to embed dialog with the public in policy-making in science, which was notably spurred in the UK, around 2000. Further to this, upstream dialog projects between scientists and the public took place in the following years across Europe, mainly with the support of the European Commission[39].

Despite 30 years of participatory TA in Europe, issues such as "why participate", "how to participate" and in order to reach "what quality" are not yet solved, which arguably weakens the legitimacy of participation and calls into question the number of citizens to be engaged, their selection (demographically or legally representative), the way to ensure the participation of big groups and the legitimacy of the participants (Pellé and

38 See footnote 51.

39 Important projects for innovation and ethical implications have been supported by Frameworks 6 and 7 of the European Commission; see Sykes and Macnaghten (2013, p. 90).

Reber, 2016). In view of all these questions related to the engagement of citizens in technological controversies, Pellé and Reber (2016) conclude that socio-political experiences in the evaluation of controversial technologies have been implemented over three decades without dealing with these unanswered questions.

Four main criteria can be identified in participatory TA (Reber, 2005): public acceptance (encompassing criteria such as representativeness, independence, early engagement, impact and transparency); deliberative democracy requirements; reflection of procedural justice and fairness; and ethics of discourse. The combination of all these criteria can vary significantly according to the importance attached to each in decision-making and in normative choices. Among other forms of TA, participatory TA systematizes the involvement of diverse social actors, stakeholders and citizens, along with technical scientists and experts, covering a variety of methods such as consensus conferences, focus groups and scenario workshops. With a varying focus on stakeholders or the public, participatory TA relies essentially on inclusion, its diversity and coherence in the process up to decision-making. Participatory TA has emerged as socio-political experimentations on emerging science/technology features and controversies, based on moral justifications intervening in the course of a public evaluation, with a varying degree of normativity (Reber, 2005). Considering risks are being tackled more effectively through widened participation, participatory TA defends the diversity of views and aims at defending all affected persons[40]. Furthermore, the complexity of decision-making in a widened participatory context implies the need to deal with the pluralism of values and disagreement:

> In general, the handling of inter-personal problems has to balance respecting the individual person – even if he/she acts in a problematic way – and the need of the majority. This may be done by using the management's right to set certain rules (for example to impose facilitation upon the group) or, quite the opposite, to reject setting specific rules by pushing the participants to take action themselves. Both sides of the balance may be seen as actions that respect the discourse ethical claims

40 However, the concern to widen the access to affected publics tends to outweigh the technical part of participatory TA. See Reber (2005, pp. 812, 825).

> for the process, as long as they are exerted with openness and transparency. (Klüver et al., 2000, p. 101)

This entails the difficulty of obtaining a clear view on the way to address the moral dimension and how to tackle disagreements in assessments through a pluralist position. While participatory TA aims at expanding to norms and values and reason-giving, the difficulty in addressing the moral dimension remains as a methodological pitfall (Reber, 2005). Therefore, it is of utmost importance that justifications and decision-making reach the same transparency as that of scientific evidence and their public restitution (Reber, 2005). Furthermore, the literature focuses also on two weaknesses that are common to participatory TA: on a pragmatic level, improving the quality of decisions requires having a widened spectrum of knowledge, experience and expertise, which is challenged by the unequal distribution among society members and, hence, their limited influence, which entails the need of enhanced access (Slocum, 2003). On a normative level, there is uncertainty regarding the way to render the decision-making process more democratic, as issues confront social norms that are either conflicting or absent, while the plurality of norms and interests complexifies the process of equal representation (Slocum, 2003).

1.3.3. *From broad participatory practices to deliberation*

Expanding the idea of public participation – through iterative processes and fair representation concerns – leads to deliberative models. Participation and deliberation are distinct notions and do not overlap, although it could be argued that the latter is a form of participation where dialog, engagement and justifications are developed to respond to essential democratic requisites, reflecting the theory of deliberative democracy. This is furthered by inter-institutional deliberation models, where institutions are responsible and in charge of such processes: public participation in policy-making is designed to facilitate a dialog with science and technology in the public sphere, in order to elicit public will (Van Lente et al., 2017). As opposed to the election-centered model, the deliberative model is talk-centered and wants ordinary citizens to deliberate issues on a regular basis (Steiner, 2012). Dialog draws from the knowledge of the public and is supposed to shed light on the stakes of the issue – from any perspective, for example political, economic, ethical or ontological – and to widen the assumptions of scientists and policymakers (Sykes and Macnaghten, 2013). In addition to its impact on governance processes, dialog can also help funders to explore how to increase the social benefit of their R&I processes.

Deliberation refers to the conduct of debate and discussion to produce "reasonable, well-informed opinions in which participants are willing to revise preferences in light of discussions, new information, claims made by fellow participants" (Chambers, 2003, p. 309, quoted in Fagotto and Fung, 2014, p. 9). Similarly, *public deliberation* can be defined as "a form of government in which free and equal citizens (and their representatives), justify decisions in a process in which they give one another reasons that are mutually acceptable and generally accessible" (Gutman and Thompson, 2004, p. 7). In deliberative democracy, the systematic involvement of public deliberation in the process of decision-making defends a high level of participation, as opposed to the traditional democratic theory, where the voting process is a sufficient source of legitimacy. Through consensus decision-making and majority rule, deliberative democracy defends public deliberation as a core element that strengthens the fabric of democracy (Fagotto and Fung, 2014), by correcting democratic deficits in policy-making (Fung, 2008) and aiming at providing "the most justifiable conception for dealing with moral disagreement in politics" (Gutman and Thompson, 2004, p. 10) – disagreement being both a condition and a challenge in deliberation (Esterling et al., 2015).

A number of successful practices in the field of deliberation have been underlined, such as good dialogic practices, the importance of hybrid forums[41] and pitfalls to avoid when conducting dialog (Sykes and Macnaghten, 2013). Deliberation can bring several positive effects such as a change in citizens' views through discussion, knowledge sharing, deepening of civic engagement, participation in public life and contribution to policy implementation. Also, moral values related to deliberative democracy are considered to strengthen the democratic fabric of our societies as greater participation contributes to fostering citizenship (Gutman and Thompson, 2004).

While deliberation could arguably be considered a key dimension in RRI, the very nature of its process and its mode remains sometimes undefined. The two fundamental questions that arise are "what" and "who": what form of public opinion is being assessed and whose opinion it is that is being

41 In order to avoid the separation between lay values and science.

assessed (Fishkin, 2009), generating a set of variables whose combination results in several methods of public consultation. Thus, before engaging in a deliberative process, some questions have to be clarified, such as which participants should be included or excluded, the nature of the decision-making method (e.g. majority, unanimity and veto right) and the way final conclusions are dealt with (Pellé and Reber, 2016).

The form of public opinion that is targeted can also vary (raw public opinion/deliberative public opinion) (Fishkin, 2009). Addressing the issue of representation and the impossible ideal of direct democracy both for practical (large number of citizens) and ethical reasons (incapacity to reach the best laws/justifications), deliberative democracy can be anchored on random sampling (Fishkin, 2009) or on democratically elected representatives of citizens; several methods of selection can be used: self-selection, non-random sampling, random sampling or everyone (Fishkin, 2009). Deliberation per se is arguably difficult to obtain from ordinary citizens, which entails the need for adapted approaches to increase the level of deliberation.

The problems and limitations of deliberative democracy derive from specific issues (Irwin, 2006) such as representativeness, with polarization seen as a possible negative antecedent for deliberation (Steiner, 2012) if the deliberation is not properly conducted[42]. Indeed, critics of deliberation have raised concerns over the possible domination of privileged members' views during deliberation processes (Fishkin, 2009). Also, the concern for public openness in deliberation arises as some theorists consider that deliberation is facilitated when it does not take place in the public eye (Steiner, 2012) – although the importance of public openness varies according to the phase of the decision process (Steiner, 2012). Another issue is the guarantee of a fair and equal representation, as the number of ordinary citizens that should be involved is debatable (Steiner, 2012).

The resulting decision-making has to respond to the reason-giving criterion by providing reasons that are accepted by free and equal persons: this moral basis implies all individuals are treated as autonomous agents taking part in the governance process, either directly or indirectly (Gutman

42 Polarization could be considered as a consequence of improper deliberation; see Steiner (2012, p. 224).

and Thompson, 2004). Further to the reason-giving requirement[43], deliberative democracy also relies on some core characteristics, such as the accessibility of the reasons given, the binding nature of the decision produced for some period of time and a dynamic process (the dialog remains open and evolving) (Gutman and Thompson, 2004). A variety of procedures and rules ensure the effectiveness of deliberative processes, such as the transparent and responsive relationship between citizens and their representatives, both being expected to justify the reasons and decisions (Gutman and Thompson, 2004). Also, deliberation has to take place in public, and deliberative justifications must be understandable and clear to those to whom they are addressed (Gutman and Thompson, 2004). When citizens rely on experts, which frequently occurs, the reasons or the bases of the reasons should still be accessible.

Although the outcomes of deliberation aim at reaching the common good, there is disagreement over consensual or pluralist results[44]. Advocates of pluralism argue that it might be "more charitable and more realistic than the pursuit of the comprehensive common good that consensus democrats favor" (Gutman and Thompson, 2004, p. 29). However, deliberative democracy does not in itself indicate a unique method for bringing deliberation to a justified conclusion (Gutman and Thompson, 2004), and, furthermore, can "rely on other procedures, most notably voting, which in themselves are not deliberative" (Gutman and Thompson, 2004, p. 18). Also, deliberative democracy can contradict the idea that what the majority decides is right, if for instance, a minority is deprived of a basic liberty (Gutman and Thompson, 2004). Indeed, as opposed to purely procedural forms of deliberative democracy, it can be argued that "the proper conception of deliberative democracy goes beyond process" and even sometimes that justice should be prioritized over deliberation (Gutman and Thompson, 2004, pp. 40–41). If public deliberation instances are usually born from the initiative and energies of civic organizations and entrepreneurs, their existence through time requires institutional support by politicians and

43 Usually opposed to aggregative social choice procedures and their voting processes, deliberation relies on reason-giving, which is its most important characteristic; see Fagotto and Fung (2014, p. 9).

44 It should be noted that philosophers in the footsteps of Mouffe and Young have criticized the aim of consensus, arguing it is both unrealistic and undesirable as an outcome, and thus the aim of deliberation could be to map the divergence of opinions. See Mouffe (1999, pp. 745–758) and Young (1990).

decision-makers. With the support of local capacities[45], public deliberation can acquire a social or political embeddedness, which is the regular habit of deliberation in the community's political institutions and social practices, on an iterative basis.

1.4. The diverse field of participation

1.4.1. *Defining participatory activities: types and timelines*

Even though citizen participation and public participation are commonly perceived as legitimate processes that facilitate an open relationship between science and society in accordance with RRI principles, the methods and outcomes of such processes are not self-evident or consistent. Moreover, participatory processes are not all integrated into R&I processes at the same stage, while participants from civil society can be individuals or groups, end-users or people affected by a certain issue, as well as a plethora of other categories, each having a different influence in an R&I process.

In the governance of science and technology, expert analysis is increasingly making space for new frameworks that foster engagement with stakeholders and/or citizens and/or the public, in order to reach more inclusive, discursive, deliberative, pluralistic, reflexive, participatory approaches (Stirling, 2005). However, while policy-making embraces more "participation", there is not a single perspective on participation, and the multiplicity of its understandings shows that in and of itself "participation" is an umbrella term, which has to respond to normative grounds and precise criteria, in order to blend concrete innovative frameworks. The wide spectrum of participatory practices is first analyzed through the complexity of its meanings and related practices, in order to define who the participants are, the new ways to include citizens and stakeholders in R&I and what these publics can bring to innovation processes.

Broadly considered, participation is a category that does not indicate the actors, the nature of the process chosen and the outcomes. Should it integrate the general public or stakeholders – and who in this latter category? Does it intervene before, during or after an R&I process and how? What are the expected outcomes and how are they being taken care of? Such questions

45 Embeddedness requires a public deliberation that is iterative and that is anchored in community-based or governmental organizations; see Fagotto and Fung (2014, pp. 13–14, 19).

need to be answered in light of the innovation modes and current challenges in R&I. Even though the legitimacy of participation in R&I tends to be self-evident, a lack of definition of the proper process may lead to poor forms of participatory practices.

Considering participation in R&I from a general point of view, the first questions that arise are: with what purpose, how and, consequently, when does participation occur? As a generic term, participation can be circumscribed to TA, which is still a wide category, while it can also refer to other dimensions. This complexity is also reflected across the related terms defining participation such as "dialog" or "engagement", "involvement" and "consultation", indicating public talk as a flexible construction[46]. Depending on their type, participatory processes can fulfill several roles, including the identification of policies/programs, the identification of undesired consequences, the improvement of quality/transparency (thanks to assessment) as well as knowledge sharing, and leverage on the impact – among others.

Stemming from a context of loss of public trust, a democratic deficit particularly salient in the current science–society relationship, participation springs up as a remedy. Through the multiple features of participatory approaches, the prevailing requirements to address transparency and openness are common drives aiming at convincing the public to trust decision-making, thus restoring legitimacy (Irwin, 2006). Parliamentary TA introduced a participative turn in TA, bringing public participation to the forefront, following the lead of, notably, Denmark and its consensus conferences[47], borrowing from the medical consensus conferences (Reber, 2006), which offer a public mediation between scientific expertise and public policies (Van Lente et al., 2017). While the Danish model focused on public debate, the Netherlands[48] developed another model based on co-production, where deliberative forums spurred interactions between mini-publics through hybrid deliberation (Van Lente et al., 2017). Regardless of their differences, these various models convey the same set of issues around the broad notion of participation.

46 In the sense that these terms are "open to variable interpretations"; see Irwin (2006, p. 314).

47 Invented by the Danish Board of Technology (DBT).

48 The Netherlands Office of Technology Assessment (later renamed the Rathenau Institute).

As opposed to representation, participation raises the issue of the involvement of new audiences in innovation processes, although the level of activity and the nature of these external audiences are yet to be defined. The increased attention on the involvement of the public in the affairs and decisions of policy-setting bodies has emerged as an international trend, commonly identified as "public participation", rephrased as "public engagement" (Rowe and Frewer, 2005). At the same time, the number of processes, techniques and mechanisms to enact involvement has also grown (Rowe and Frewer, 2005).

Widely understood as involvement, "participation" is a wide category that can refer to various forms, contexts, types of participants, requirements and aims: the related mechanisms that can be leveraged to reach effective involvement are also diverse (Rowe and Frewer, 2005). Citizens and stakeholders can variously connect to R&I processes in terms of contribution and proximity – either internally or externally to the R&I process[49]. Under this bipartite distinction, three categories of participatory activities are outlined:

– citizen and stakeholder engagement *in innovation projects*;

– citizen and stakeholder engagement *within agency processes*;

– citizen and stakeholder engagement *in evaluation processes*.

Within these categories, several types of activities can be outlined, helping to separate the different outcomes of activities involving the participation of citizens:

– *general consultation*: processes of engagement with any group of citizens or stakeholders, in order to obtain feedback on policies and programs;

– *evaluation (of projects and programs)*: evaluation of project proposals (i.e. the ethical and scientific evaluation) as part of the selection process intervening in funding schemes, interim and *ex post* evaluation for projects and programs that receive funding and program evaluation[50];

49 This distinction may have direct consequences in the way participants should be treated (e.g. in medical research on human participants, research on vulnerable people is often only allowed if they directly benefit from it).

50 Participation in assessment of science and technology has been developed for more than 30 years through technology assessment and contributes to agreements on substantial investments

– *information/knowledge sharing*: dissemination of information on R&I contents, outputs and scientific knowledge. Also one-way, like the communication of funding activities, knowledge sharing relates to the dissemination of scientific content only. This mostly relates to the one-way engagement toward any type of public and the sharing of information pertaining to the communication of funding activities (funding portfolio) and/or funded activities (funded research/development/innovation and its results);

– *monitoring*: systematic follow-up of funded projects in the context of funding schemes, as an activity usually carried out internally (but which can be supported by external experts: e.g. when doing interim or final reviews), although *ex post* monitoring of results can involve other actors, in addition to the involvement (feedback) of programs' beneficiaries;

– *program design*: in the context of R&I, program design refers to the identification of program objectives and of R&I priorities, resulting in the definition of funding opportunities;

– *social impact evaluation*: activities that aim at measuring the societal impact of R&I processes, including the development of indicators and the monitoring allowing such measurement;

– *project execution*: involvement of external participants[51] in the execution of an R&I program/project.

Other categorizations can be found in the literature, for instance a tripartite distinction based on the objectives of the policy process, which can involve: transmitting information (unidirectional), consultation (bi-directional) or active participation (when all involved parties contribute to the issue to a certain extent) (Slocum, 2003). Although the latter seems to better reflect the idea of "participation", all forms are considered as part of participatory approaches, the common denominator being the involvement of

---

and technical options; better responses to social needs; agenda-setting; mapping of scientific controversies; more interactive surveys. The benefits of participation in evaluation processes are mainly that by expanding the viewpoint of experts with perspectives of other epistemic communities, it can help reach a more comprehensive view of a technological innovation and its impacts on society. Assessment can refer to distinct processes, depending on the stages of innovation processes: *ex ante* evaluation, *interim* evaluation and final and *ex post* evaluation. Each stage has a different array of implications, while evaluation can also be considered throughout the whole timeline, as *in itenere* assessment (during all phases).

51 The types of participants are explored in the next section.

"the public" (Slocum, 2003). According to this categorization, participatory approaches take place in three distinctive areas: evaluation, planning and implementation.

In a general sense, public participation is "the practice of involving members of the public in the agenda-setting, decision-making, and policy-forming activities of organizations/institutions responsible for policy development" (Rowe and Frewer, 2005, p. 253). With varying levels and ways of participation, the broad category of public participation could be opposed to non-participation situations related to traditional models of governance, where along with nominated experts elected policymakers set policy without reaching out to the public (Rowe and Frewer, 2005). According to the level of involvement and the goals, certain forms may not be considered as participatory.

1.4.2. *Defining participants: types of actors and visions of engagement*

Commonly identified as an "inclusion" mechanism, participation can embrace various kinds of participants[52], including stakeholders, citizens, the general public (lay people) and civil society – often as composite publics. Participation actors are another blurred area in the definition of participation, usually identified as "citizens" or "stakeholders", which are not overlapping terms, nor do they cover the whole array of potential participants. If participatory processes in innovation involve mainly citizens, with the aim of bridging science and society, other actors can be identified under the category of "stakeholders". As the variables in participation are multiple, so are the models, depending on who participates, what the underlying intention is, who conducts the procedure, what the distribution of (technical) expertise is and if there really is a choice addressed to participants (Rowe and Frewer, 2005).

Participants can be divided into internal or external actors (in relation to the R&I process), involved in activities happening either before, during or after R&I processes. According to the field – for example funding agencies – participants are perceived as different entities and can embrace a scope that can either be specific or wide. Generally seen at the intersection of science and society, participants usually refer to multi-actor approaches involving

52 In addition to the usual participants (experts, scientists).

societal engagement, stakeholder engagement, citizen dialog and any interaction that brings guarantees that the social impact of new technologies and innovations can be considered in R&I activities. Participation also relates to the wide category of participatory democracy, which is fundamentally a means to achieve the very ideal of self-government (Fung, 2003). Throughout the various participatory approaches, the common denominator is the involvement of "the public", which "can be average citizens, the stakeholders of a particular project or policy, experts and even members of government and private industry" (Slocum, 2003, p. 9). On the contrary, a restrictive approach limited to participatory processes in funding schemes refers precisely to evaluation processes and ethical appraisals.

Some general divides can be outlined[53]:

– *participation orchestrated by funding/non-funding entities*: in ethics review (appraisal) schemes, experts and researchers are participants taking part in review processes, ensuring the proper ethical/scientific evaluation of activities according to ethical requirements. Also, some beneficiaries in funded R&I projects can be considered as "citizens participants" or "participants in novel modes of interaction", considering that their role is to implement the funded proposals. The compliance with ethical requests is a current feature in participatory processes orchestrated by public funding entities;

– *participants as citizens/stakeholders*: community-based and creative methods to involve citizens differ from stakeholders' involvement, being specifically focused on categories of citizens that may have an interest in R&I activities. As a category, "citizens" is broader than "stakeholders" and can ultimately embrace all types of participants in their belonging to a given society. "Citizens" can also be seen as a category opposed to "stakeholders" as it implies an action toward the common good, ultimately raising the question of the societal purpose or implication of an R&I activity;

– *participation in creation or implementation (operational working processes)/in evaluation (assessment)*: participants involved in the operational working process of R&I activities differ from those involved in evaluation processes. Contributing to programs' design or implementation involves a different engagement than participation in evaluation, where the

53 The following distinctions concern different dimensions: these are not disjunct and can be combined.

contribution focuses on compliance with ethical frameworks, regulations and procedures.

Participants in R&I can generally be outlined through the two main categories of citizens and stakeholders: the first one addresses the concern of public engagement, whereas the second one ensures all concerned parties are involved in assessment and decision-making, resulting in either policies or R&I products. Both citizens – as the general public – and stakeholders can contribute to fostering responsible innovation, helping to incorporate relevant ethical and societal aspects into innovation practices and to achieve desirable goals, by discussing, assessing directions and consequences and setting priorities (see Blok et al. (2015)).

Stakeholders can be defined as individuals, groups or organizations who can affect or be affected by an organization's activities, although a distinction can be made "between economic stakeholders like employees and suppliers, and non-economic stakeholders like NGOs and research institutes" (Blok et al., 2015, p. 149). Stakeholders represent shared interests, while they also respond to the notion of inclusion and sustainability. As an accountability mechanism, stakeholder engagement targets a purpose to achieve agreed outcomes (AccountAbility, 2015)[54]. On the level of inclusivity, participation of stakeholders aims at achieving accountable and strategic responses to sustainability (AccountAbility, 2015). Stakeholder engagement contributes to responsible innovation by bringing transparency, interaction, responsiveness and co-responsibility (Blok et al., 2015). On a wider scale, stakeholders also complexify controversies and debates in innovation processes, as they operate on the level of the representation of interests.

Throughout the categories of participants, the most meaningful variable might be the level of responsibility that is assigned to them, according to the type and mechanism of participation. Identifying new ways to include citizens and stakeholders in R&I requires specifying the context and the scope.

54 The AA1000SES (2015) is a generally applicable framework for the assessment, design, implementation and communication of quality stakeholder engagement.

A more detailed categorization would allow a distinction of roles according to the type of participation:

– *citizens*: ordinary people, defined as the general public[55], lay people and citizens as persons (or collectives) with civic expectations[56];

– *end-users/consumers*: a specific category distinct from citizens (sometimes with overlaps), referring to beneficiaries of the end product (including solutions and services) of the R&I process. This category specifically refers to the impact and the uses of the products of R&I activities in society;

– *experts*: individuals enrolled as (internal or external) experts in RFO activities, but at the individual level and not as part of any of the other categories. Therefore, lay experts for instance can be included in this category, if their involvement mainly values their expertise (as citizens) and if this category prevails over others (e.g. "citizens"). Experts can be individuals with any sectoral expertise (e.g. with a background in medicine, psychology, sociology, philosophy, among others);

– *private bodies' representatives*: representatives of small- and medium-sized enterprises (SMEs) or large companies (including those conducting research), considered as beneficiaries (actors who implement the funded proposals) or non-beneficiaries of funding schemes and involved in some way in RFO activities as participants, participating on behalf of their company. This category also comprises social entrepreneurs;

– *public bodies' representatives*: also understood as governmental institutions, this category encompasses national public bodies (e.g. government departments or ministries, excluding academic research institutions) that might be involved in the management of RFO activities and projects or that might operate as funding agencies. Public bodies relate to ethics through ethical compliance mainly, ensuring ethics requirements are met in grant agreements signed with the beneficiaries at a national level;

– *organized civil society representatives*: civil society organizations have different levels of knowledge and leverage compared to individual citizens.

---

55 Including citizen scientists and open science movements.

56 In this category, the main boundaries for the "citizens" category are anchored in the citizens/stakeholders distinction. Also, since end-users can be categorized as citizens as well, this distinction serves to underline the general dimension of involvement, referring to the broader sense of "public participation". This category may also relate to a broad participatory approach.

They may defend interests, often professional interests (trade unions), causes (e.g. animals and environmental issues) or rights (e.g. minorities and women).

– *research participants*: human participants who are part of the subject of research (for instance, in medical sciences, the public and patients' involvement).

– *scientists and research institutions' representatives*: individual scientists (any scientific or technical field) or representatives of research institutions involved in RFO processes either for their individual expertise or for their affiliation to a research institution, but speaking for themselves only. In cases when they are invited on behalf of a private company conducting research, these participants should be identified as part of the "private bodies" category. This category also includes researchers as physical persons working for a legal entity (SMEs, universities, research institutes) conducting research and to whom public or private funding has been granted to do so. Participants under this category can be considered as "experts".

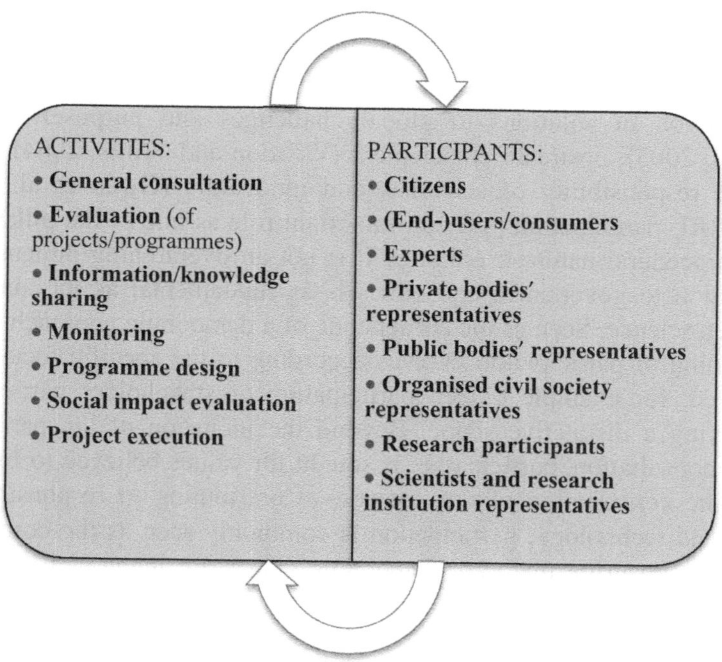

Figure 1.1. *Types of activities and participants in R&I participatory processes*

Some categories can overlap, according to the nature of the participation, as is the case for "citizens", who may be, for instance, directly involved as research participants. In assessment (evaluation), several types of experts/evaluators can be identified, depending on the proximity with the project: self-assessment, internal, external (neutral) and evaluation by beneficiaries (or users)[57]. Regarding knowledge sharing, both researchers and citizens can be involved, while both are also involved in other categories, since researchers participate in R&I activities and citizens in consultations on policies. Firms and public bodies are mainly represented in the implementation of projects, while citizens can also be represented through civil society organizations and not necessarily through individual direct involvement.

1.4.3. *Finding value in participation*

Participatory practices introduce new challenges in RRI, requiring an additional layer of scrutiny on the review of ethical issues, whether formalized (as in ethical reviews) or not. At the same time, participatory practices are a core element of RRI: the RRI concept has emerged as an extension of the science in society discourse (Khan et al., 2016) about the co-production of solutions to global challenges and purposeful science (Jasanoff, 2005), upstream engagement (Wilsdon and Willis, 2004) and the reflexive responsibility of scientists and innovators (Owen et al., 2012). Within RRI, participation plays an important role as one of the pillars/keys, with a procedural nature[58]: although it is not an overarching dimension, its contribution to governance and ethics is as fundamental as its connection with open science. Seen as the cornerstone of a democratic approach to R&I, the meaning of participation evolves according to the specific focus that is considered, for example citizen participation or stakeholder participation, each having a distinctive scope. Beyond the inclusion of lay participants through consultation, participation is sought for values believed to foster the democratic construction of our societies. Contributing to responsibility in science and technology, participation is commonly seen as the corollary of democracy, ensuring that core democratic values are integrated throughout

57 Toward participation in the design and implementation of follow-up/monitoring.

58 It could be argued that participation, governance and openness are of a rather procedural nature, focusing on cooperation, whereas gender equality, scientific education and ethics are more of a substantial nature. See Pellé and Reber (2016, Chapter 2).

time and frameworks. At the core of the science–society relationship, participatory processes play a vital role in supporting democracy, even though democracy is not always mentioned in TA or RRI; see for instance Pellé and Reber (2016).

It is however acknowledged that complex public problems can be tackled efficiently by opening channels of participation, inviting citizens and stakeholders in public decision-making (Fung, 2008), especially considering these voices can be at odds with expert and scientific positions (Sykes and Macnaghten, 2013). Depending on the conditions and the nature of participation, the value of this latter varies significantly: participation practices may lead to unfair concentration of power in the hands of a privileged, educated elite and would undermine the interests of disadvantaged groups who have not been able to engage in participation. Considering their potential added value, participatory practices are considered to improve the quality of decisions, whereas on a normative level it can be argued that they ensure decision-making processes are more democratic (Slocum, 2003).

The value to be found in participatory practices also lies in the capacity to introduce dialog and debate in an arena that 50 years ago was rather undisputable. Acknowledging that science and technology can have unforeseen effects – and that they are no longer solely positive, neither for humans nor the environment – appeals for more dialog. Participation can help tackle this issue, especially taking into consideration that "[science] is also part of the contemporary social and economic order, and closely tied to processes of industrialism and consumerism" (Sykes and Macnaghten, 2013, pp. 86–87).

The question of the ethical grounds and the legitimacy of participatory processes points to the need for the contribution of citizens and stakeholders to R&I. Participatory processes have several benefits, in decision-making, in R&I processes themselves and in priority setting. These benefits could be summarized as follows:

– *impact measurement*: insight into undesired consequences, as well as on the degree to which R&I policies are aligned with societal needs, through the involvement of new actors, such as end-users and co-creators in the R&I process;

– *transparency and accountability*: there is an increase in transparency and accountability levels in R&I processes as a result of evaluation processes;

– *allocation of funding*: participation can help in priority identification and the justified allocation of resources through competitive funding;

– *science–society dialog*: participation can also consist of a consultation or dissemination endeavor to share knowledge and merge the gap between citizens and scientific processes.

These processes inform and help decision-making, by providing leverage on the impact (bringing R&I policy aligned with societal needs), avoiding undesired consequences, setting priorities (and fostering competitive funding) and enhancing the transparency, quality and accountability of R&I processes. A thoughtful participation can also avoid conflicts of interests (which can occur both from internal or external stakeholders) by anticipating them and reasoning in terms of common good or finding compromises.

Could there be an ethical position advocating non-participation in R&I? Traditionally left in the hands of experts and scientists, technical issues cannot always be embedded in participatory processes (Rowe and Frewer, 2000). The traditional divide between experts and the lay public derives from the belief that decisions regarding technical issues should be handled by experts and scientists alone: the extended view of participatory processes is counter-intuitive in the context of a traditional approach to science and technology governance. Furthermore, without participation, there may be a minimal representative policy process (Fung, 2008), where the representation of citizens is seen in the translation of citizens' interests into policies by politicians, and where accountability intervenes in the outcomes only through the voting process.

While fostering dialog can be considered as a key benefit of participation, debates can either be direct, through deliberation, or indirect, by allowing an inclusive approach on certain matters. The tendency toward the professionalization of public engagement has shaped dialog differently over the past decade, increasing the divide between academic social scientists and dialog practitioners, leading some to criticize a "public engagement industry whose commercial interests can compromise democratic ideals of participation" (see Chilvers, 2010, quoted in Sykes and Macnaghten, 2013, p. 100). This issue can be tackled by intensifying the process of dialog, in deliberative processes for instance, or more generally by ensuring participation takes place at all stages of innovation (Sykes and Macnaghten, 2013), with enough transparency and feedback from lay people, whose viewpoint should gain legitimacy. Reinvesting participation this way aims at a process where publics and issues emerge in a pluralist, inclusive and interdisciplinary way (Sykes and Macnaghten, 2013), bringing several criteria: intensity, openness, quality and responsiveness.

The main benefits of participatory science and innovation spread out in two directions: knowledge added value (cost benefits, speed of delivery, diversity of skills mobilized and diversified observations) and societal or social added value (in education, in fostering the science–society relationship, in citizen empowerment and in problem solving and skill building) (see Houllier and Merilhou-Goudard (2016)).

1.4.4. *Models of participation*

In the absence of regulations on participatory practices in R&I, the literature on the subject reveals the compartmentalization of the different mechanisms and, at the same time, the complexity of a global distinction of participation modes. Across the scientific literature on the subject, categorizations reveal various viewpoints and criteria in terms of processes and objectives.

1.4.4.1. *Value-based categorization in view of empowerment*

One of the classic texts on participation is Arnstein's (1969) so-called "ladder of participation", where the bottom represents low active involvement (considered as non-participation) and the top represents high engagement through eight ascending rungs: "manipulation", "therapy", "informing", "consultation", "placation", "partnership", "delegated power" and, at the upper level, "citizen control".

The two lower ranks, termed "manipulation" and "therapy", are identified as "non-participation", as these forms do not enable citizens in planning or conducting programs, but enable powerholders to "educate" or "cure" the participants. Going up in this ladder, the next degrees are ranked as degrees of "tokenism" as they allow the underrepresented categories to have a voice. While "informing" and "consultation" levels allow citizens to be informed and be heard, at the same time their views are not transferred to higher levels: "when participation is restricted to these levels, there is no follow through, no 'muscle', hence no assurance of changing the status quo" (Arnstein, 1969, p. 217). Also, in the same category of "tokenism", the mechanism of "placation" allows citizens to advise, although the separation from powerholders is still present.

Upper levels are grouped as "degrees of citizen power", as they are considered to bring the empowerment of citizens, either through "partnerships" with traditional powerholders or at higher levels, as "delegated power"; while at the highest level of empowerment is "citizen control". In the two topmost rungs, "have-not citizens obtain the majority of decision-making seats, or full managerial power". Criticism over this ladder model of participation has raised the issue of the underlying judgment on what type of participation is best, in an ascendant path (Davidson, 1998; Rowe and Frewer, 2000; Rowe and Frewer, 2005; Reed, 2008).

1.4.4.2. *Contextual models of participation*

As an alternative to the ladder, the wheel of participation (Davidson, 1998) states that four categories of participation (*inform*, *consult*, *participate* and *empower*), each of them different in intensity, can be equally appropriate, depending on the context. In this model, the term "participation" can refer both to the whole wheel or to categories within the wheel. When participation refers to the whole wheel in a comprehensive understanding of participation, it covers all the means by which affected publics take part in policy formulation or implementation[59].

Another model of participation that distinguishes contexts of use is the one rooted in the distinction of activities. Rowe and Frewer (2000, 2005) propose to rephrase the "disparate area of public participation as public engagement", in which they distinguish three significantly different activities: public communication, public consultation and public participation (Rowe and Frewer, 2005, p. 284). The three concepts have been differentiated according to the nature and flow of information between the organizers of the participatory activity and the participants.

Public communication. Information is conveyed from the sponsors of the initiative to the public. In this dimension, information flow is one-way, as there is no involvement of the public in the sense that public feedback is not required or specifically sought. When the public attempts to provide information, there are no mechanisms specified *a priori* to deal with this at

59 This view is also consistent with Richardson's definition of participation (Richardson 1983).

any level beyond, perhaps, simply recording the information. Effectiveness is conceived in terms of “maximizing the relevant information from the sponsor and efficiently transferring it (with minimal information loss) to the maximum number of the relevant population, with the efficient processing of that information by the receivers (the public/participants)” (Rowe and Frewer, 2005, p. 263).

Public consultation. Information is conveyed from members of the public to the sponsors of the initiative, following a process initiated by the sponsor. The information elicited from the public is believed to represent currently held opinions on the topic in question. The overall process seeks effectiveness, by “maximizing the relevant information from the maximum number of the relevant population and efficiently transferring it […] to the sponsor” (Rowe and Frewer, 2005, p. 263);

Public participation. Information is exchanged between members of the public and the sponsors. That is, there is some degree of dialog in the process that takes place (usually in a group setting), which may involve representatives of both parties in different proportions (depending on the mechanism concerned) or, indeed, only representatives of the public who receive additional information from the sponsors prior to responding. Rather than simple, raw opinions being conveyed to the sponsors, the act of dialog and negotiation serves to transform opinions in the members of both parties (sponsors and public participants). Effectiveness is conceived in terms of “maximizing the relevant information from the maximum number of all relevant sources and transferring it […] [with] the *combining* of it into an accurate composite” (Rowe and Frewer, 2005, p. 263).

This three-tier classification has been further detailed into a typology revealing four classes of communication mechanisms, six classes of consultation mechanisms and four classes of participation mechanisms (Rowe and Frewer, 2005).

A final interesting approach in this classification of participatory mechanisms is one that, instead of relying on the type of activities, focuses on the variability in the characteristics of responsible stakeholder engagement, in regard to the different phases of innovation. Following this approach, some key characteristics allow the following categorizations (Blok et al., 2015): “transparency”, “responsiveness”, “interaction” and “co-responsibility”.

Value-based categorizations in view of empowerment and contextual categorizations of participation show the extent of discrepancies between the various mechanisms termed "participation", although they have little in common in terms of publics, outcomes, timelines and relationships with responsible innovation goals.

1.5. Conclusion

In the course of this chapter, several features have been separately explored, on the one hand, in the over-regulated field of ethics of R&I, and, on the other hand, in the under-regulated field of participatory practices. From the viewpoint of institutionalized processes in R&I governance, it seems difficult to draw a general picture of ethical participation, since both sides do not automatically converge. In the case of ethical considerations in R&I, both the theory and the regulations are quite advanced, delineating several sectoral and general guidelines to be followed, usually as part of binding tools (regulations) and frameworks, in particular in the case of public funding schemes emanating from the European Union's R&I programs. Quite different, if not opposed, participatory practices represent a variety of provisions, configurations, aims and objectives, which defy the possibility of a broad overarching approach. Several taxonomies and categories have been outlined in gray and scientific literature, without any concrete connection to the ethics of R&I. As a poorly regulated field, it is also a messy landscape of different practices, led without the existence of proper guidance. The taxonomies in participatory practices that have been presented are further connected to the ethics of R&I in the following chapter, in the attempt to better explore what ethical participation could bring in terms of added value and the legitimacy of R&I processes.

2

Regulations on Ethics and Participation

Following the previous analysis of the framing of ethics and participation in R&I, focusing on their common understanding and theoretical connections, this chapter delves into the regulatory aspects and the current needs as discussed in the existing scientific and gray literature. The first section (section 2.1) focuses on the main normative instruments concerning the ethics of innovation at the European level. The intricacies and boundaries of current European regulations are then explored in relation to ethics and participation (section 2.2). While there is no consistent regulation supporting participatory practices, soft law offers interesting pillars (section 2.3) worth considering as a basis for an effective shaping of participatory practices. Considering the current regulatory limitations, the way to embrace ethics and participation R&I can be explored beyond regulations (section 2.4), through the gaps perceived on behalf of R&I institutions. This chapter also builds on gray literature, policy recommendations and analysis of existing gaps in the institutionalization process and existing frameworks from regulatory bodies.

2.1. Policies and regulations on the ethics of innovation

2.1.1. *Main regulations on research ethics and research integrity*

The main assumption of this study regarding European regulations on ethics and participatory practices is that, at first glance, the landscape is rather blurred, with a silo structure of separate regulations: on the one hand, the ethics of R&I and, on the other hand, participation-related regulations. These latter do not embrace the whole spectrum of participatory practices and they do not necessarily cover the field of ethics of R&I: these two aspects add a layer of complexity in the delimitation of these regulations. An initial look at the main regulations on research ethics and research integrity helps in better observing, as a second step, the main shortcomings of participation-related regulations.

European regulations extend the ethical issues taken into consideration in European funding procedures and policy, and they also complete a list of former international regulations on research ethics and research integrity. Regulations intervene in the the following areas of concern:

> Applying established ethical principles and legislation to research involving children, patients, vulnerable populations; use of human embryonic stem cells; privacy and data protection issues; research on animals and non-human primates. [They also include] established principles of research integrity (data fabrication, falsification, plagiarism or other research misconduct). (Owen and Pansera, 2019, p. 38)

Two founding legal texts on research ethics address the issue of – and set the basis for – the protection of research participants:

– the Nuremberg Code, formulated in 1947 in Nuremberg (Germany) by the American judges of the Nuremberg Tribunal, in the aftermath of World War II and the revelations of German and Japanese medical experimentation on human subjects. It establishes the notion of free and informed consent for all human participants in biomedical research and sets 10 fundamental principles related to consent, proportionality and necessity as well as the right to withdraw;

– the Helsinki Declaration of 1964[1] on research ethics, which sets ethical principles for the conduct of medical research on human subjects, involving the protection of research subjects (the well-being of the individual research subject must take precedence over all other interests).

These two texts have been followed by other international normative instruments – mostly on research ethics – including the following:

– the Convention on Human Rights and Biomedicine or the Oviedo Convention, adopted by the Ministers of the Council of Europe in 1996,

1 Revised in 1975.

which sets standards[2] for all members of the Council of Europe regarding ethical issues raised by research within the framework of the protection of human rights;

– UNESCO's Universal Declaration on Bioethics and Human Rights;

– the Council for International Organizations of Medical Sciences' (CIOMS) International Ethical Guidelines for Biomedical Research Involving Human Subjects;

– the European Charter of Fundamental Rights (articles referring to research integrity and to data protection).

Many ethical issues are covered by legal instruments in the European Union; however, research ethics go beyond legal frameworks and require independent evaluation at all stages, in line with RRI. Part of this goal is covered by the ethics review procedure[3] that takes place in the assessment of projects requesting funding from the European Commission. Also, there is a strong connection between research ethics and human rights, with reciprocity and overlaps: within the European regulatory framework, research ethics is based on the commitment to human rights, a compliance that is relevant for all policy domains.

Current legal literature on the ethics of R&I is covered both by international codes of conduct and voluntary codes, at the international or European level. This literature can be classified by geographical scope[4] or thematically, some of these regulations being general or covering specific topics in the ethics of R&I.

2.1.1.1. *Research integrity*

A number of integrity codes in research have been issued at the European level: they cover the general principles in research integrity and are furthered by voluntary codes (legally non-binding) issued by several European universities. The European Code of Conduct for Research Integrity (All

---

2 Including standards for the use of the human genome and research on human embryos.

3 This procedure is explored in the next section.

4 This is, for instance, the presentation mode of the ENERI project (EU-funded database for all codes, guidelines and recommendations, both national and international, covering all topics of research ethics and research integrity): http://eneri.eu/codes-of-conduct/.

European Academies, 2017)[5], compiled by All European Academies (ALLEA) in 2011, sets the general principles for research integrity, informing national research policies in Europe and applying to research in all scientific and scholarly fields, and both in publicly funded and private research, as a framework for self-regulation[6]. The main principles (reliability, honesty, respect and accountability) are followed as good research practices, failures of which are perceived as "violations of research integrity". These principles can be applied to research funding, which in turn becomes a tool ensuring compliance with the agreed framework. However, ethical principles require careful scrutiny: accountability for instance is an upstream issue – prior to the research activity – and also a determining element to prioritize funding, and it responds either to scientific policies or even politics (societal challenges) (Pellé and Reber, 2016). Downstream of the research process, developments can bring about unexpected collateral effects or controversies[7]. Following these guidelines, several national non-binding codes of conduct have emerged, either with a sectoral focus or with a cross-cutting perspective, following the self-regulation principle of the European Code of Conduct for Research Integrity[8].

Addressing the needs of the information society, data protection is covered by general legal texts on research integrity[9] and specifically by the EU's Regulation 2016/679 – General Data Protection Regulation (GDPR). Considered by the European Commission both as a central issue for research ethics in Europe and a fundamental right, data protection is linked to autonomy and human dignity (European Commission, 2018). This main legal source regarding data protection is grounded on founding texts of the European Union: the European Charter of Fundamental Rights and the

---

5 This code of conduct was published in 2011, with a revised edition in 2017 (considered here) encompassing emerging challenges emanating from technological developments, such as open science, citizen science and social media. This code of conduct has been set as a "living document", to be revised every five years at the most in order to take account of evolving concerns.

6 As stated in its Preamble.

7 Especially in sectors such as biomedical research, genetically modified organisms or nanotechnology. See Pellé and Reber (2016, Chapter 1).

8 Such as the sectoral focus of the Code of Ethics for Estonian Scientists or the (more general) Netherlands Code of Conduct for Research Integrity.

9 The European Code of Conduct for Research Integrity includes some guidelines on "data practices and management", already included in the 2011 version (see in particular sections 1.4, 2.2.4, 2.2.5 and 2.3) and further developed in the 2017 version.

Treaty on the Functioning of the European Union. With the growing use of digital data, the connection of data with risk regulation has grown to a major concern, which the GDPR tackles by strengthening individual rights vis-à-vis data controllers – either public or private entities. Data processing operations are part of a shift toward a new model in the management of uncertainty that revolves around enforced self-regulation (Spina, 2017, p 89). Data protection is also connected to data ethics, as ethical considerations attempt to tackle the epistemic change that big data have brought in knowledge availability and dissemination, and more generally in the governance of public and private space (Spina, 2017, p. 92).

2.1.1.2. *Research ethics: the protection of research participants*

Science and society issues diverge from scientific integrity issues: the first relate to problems depending on the socio-ethical context of research, whereas the second category focuses on standards that are relevant when conducting research[10].

Even though research ethics is most developed in medical research, the general principles apply to all fields of research[11], bringing principles, such as informed consent or confidentiality, that are cross-sectoral. Already addressed by several international legal frameworks (see above) since the Code of Nuremberg, the issue of the protection of research subjects – research ethics – is mainly covered by the Belmont Report, issued in 1978 by the National Commission for the Protection of Human Subjects of Biomedical and Behavioral Research, which sets Ethical Principles and Guidelines for the Protection of Human Subjects of Research.

The objective of the Belmont Report is to provide an analytical framework serving to guide the resolution of ethical problems arising from research involving human participants through general prescriptive judgments, that is "basic ethical principles"[12], on: (1) respect; (2)

10 This publication also mentions that "there is, of course, no perfect watershed between the two categories". See the first edition of the aforementioned code of conduct (All European Academies and European Science Foundation, 2011, p. 10).

11 This section is mainly based on the publication by the European Commission Directorate-General for Research and Innovation (2013).

12 The Belmont Report identifies them as such: see part B of the Belmont Report (National Commission for the Protection of Human Subjects of Biomedical and Behavioral Research, 1979).

beneficence (avoiding harm); and (3) justice. These three main ethical principles are completed by a set of requirements in their application, which are: (1) informed consent (information, comprehension, voluntariness); (2) assessment of risks and benefits (nature and scope, systematic assessment); and (3) selection of subjects.

As a common denominator of these regulatory frameworks, risk regulation tackles scientific and technological uncertainty, while risk assessment, risk management and risk communication are all intertwined in the same precautionary position:

> risk regulation has been traditionally associated with the uncertain negative outcomes connected with the products of industrialized manufacturing, such as food, pharmaceuticals, chemicals, or with sources of energy, safety of transport, or general environmental issues. (Spina, 2017, p. 88)

Overall, current regulations on the ethics of R&I cover the ethics of participation only to the extent they relate to the involvement of research participants, but not specifically in innovation governance and innovation processes themselves. Further to regulations, such issues can be tackled by ethical processes, already operational through formalized procedures in public R&I funding.

2.1.2. *Ethics assessment: a standard in European funding programs*

Further to international and European regulatory frameworks, ethics also plays a key role in the stream of European R&I policies and the projects that they support. Ethics assessment provides moral judgment on R&I practices, projects and developments, ensuring their conformity with moral values, principles and norms: these assessments can be project-/practice-oriented, policy-oriented or related to professional conduct (Shelley-Egan et al., 2016).

Within European funding schemes in R&I, this takes shape through ethical compliance processes, which is a main pillar, although under-represented, in RRI literature (Pellé and Reber, 2016). Programs supported by the European Commission respond to the research proposal evaluation procedure, the so-called "ethics review", whose purpose is to ensure the compliance of funded research activities with fundamental ethical principles. The various Framework Programs of the European Commission represent a great proportion of public funding in Europe and their compliance with ethical principles is a key element in their operational mode. The legislation that sets the standards for each program underlines the fundamental principles that have to be considered, as well as some precautionary principles on specific fields of research[13]. Evaluation is useful feedback for decision-making, providing systematic assessment of the implementation or merit of a program, by drawing conclusions based on empirical analysis of data (direct or indirect) (Oldsman and Nexus Associates, 2014). The two main sensitive ethical issues are arguably research involving humans and the protection of personal data.

The current ethics review procedure was established in 2011[14] and applies to proposals submitted to the European Commission that have been retained with a view to funding but identified as raising ethical issues. Indeed, research proposals are evaluated both on their scientific merit and on the ethical and social impact as to whether they meet ethics requirements. At the European Commission, ethics reviews are performed by a panel of experts in two phases[15]: (1) an ethics screening and (2) a (full) ethics review (ethics assessment). The ethics review procedure intervenes in a wider timescale of project development, as follows: submission, with ethical self-assessment; assessment to determine the scientific value of the project and establish a ranking; ethical review; preparation for allocation; and project implementation (Pellé and Reber, 2016).

13 Certain fields of research cannot receive funding (e.g. human cloning) or other fields may be financed if they demonstrate compliance with licensing and control procedures (e.g. research on human embryonic stem cells).

14 The ethics review procedure was established in 2011 by the Commission Decision of 28 February 2011 amending Decision C(2008) 4617, "related to the rules for proposals submission, evaluation, selection and award procedures for indirect actions such as the Seventh Framework Programme of the European Community for research, technological development and demonstration activities (2007–2013) and under the Seventh Framework Programme of the European Atomic Energy Community (Euratom) for nuclear research and training activities (2007–2011)".

15 These two phases have the purpose of identifying projects in need of a follow-up or audit to assist the beneficiaries to deal with the ethical issues raised by their work, eventually ending in preventive and/or corrective measures.

The ethics review occurs in the intermediary step of the three-step ethics appraisal process:

– *ethics self-assessment* by the applicant;

– *ethics review* by ethics experts, before the finalization;

– *ethics checks* by ethics experts, and officers[16], for selected projects, after the signing of the grant agreement.

Early on in the development of a research project, a list of ethical issues outlined by the European Commission has to be taken into account, as ethics panels could be globally qualified as risk-averse (Karatzas, 2020). Eleven ethics issues are considered in European ethics review procedures, also known as the ethics issues "checklist"[17]:

1) human embryos/fetuses[18];

2) human beings;

3) human cells/tissues;

4) personal data[19];

5) animals[20];

---

16 Officers from the European Commission (initially) and now especially from the European research area (ERA).

17 This procedure applies to all projects that seeking fund from the European Commission. See European Commission Directorate-General for Research and Innovation (2019).

18 Embryonic stem cell (hESC) research raises quite important ethical questions on sensitive issues related to human life that have received a variety of answers in Europe according to specificities of national regulations on that topic. As a consequence, the European Commission accepts only research projects that are allowed in all European Member States This issue is completed at the international level by the UNESCO Universal Declaration on the Human Genome and Human Rights.

19 Aims at strengthening the rights of individuals, data protection and privacy issues related to the principles set in the Charter of Fundamental Rights (Official Journal of the European Union, 2010a): articles 7–8; the Treaty on the Functioning of the European Union (Official Journal of the European Union, 2010b): article 16; and the General Data Protection Regulation (GDPR) (see below in this chapter). These principles aim to guarantee the right to privacy and refer to the technical framework and security measures designed to guarantee that all personal data are safe from unforeseen, unintended or malevolent use.

20 Conducting research on animals (animal testing/research) should be compliant with three principles: *reduction* (methods that can reduce the number of animals), *replacement*

6) non-EU countries[21];

7) environment, health and safety;

8) dual use[22];

9) exclusive focus on civil applications;

10) misuse;

11) other ethics issues.

Beyond ethics compliance and its primary use as a recommendation to researchers to think about ethics while designing research protocols, the ethics review procedure acts as a powerful tool for reflexivity in the R&I process, in general. Extending it to innovation actors too, the set of ethical issues considered in ethics reviews includes significant items, whose use could be exceedingly wider than its original context of use.

Taking stock of current challenges and approaches in innovation, the ethical landscape and the regulatory schemes applying to R&I, this section has highlighted the importance of anchoring future developments on ethical (prescriptive) strands that surpass existing legal schemes and normative/ regulatory approaches. The risk of over-regulation and the weight of normative features loom on the horizon as R&I often shows the need of independency. Participation as entangled with ethics shows another way forward.

(prioritize methods that do not involve animals) and *refinement* (all methods should alleviate/minimize the potential pain for animals and enhance animal welfare).

21 When research is conducted in or in collaboration with non-EU countries, the activities must comply with the general objectives of the Union's external action and also be compliant with all European standards. In the case of developing countries or emerging economies, special measures have to be taken to ensure that the rights and interests of all participants are adequately protected and that the benefits of the research are equally shared.

22 This term refers to technology that can be used both for peaceful and military aims: the possible dual use of new technologies implies potential misuse (in the event that research activities either involve or generate materials, methods or knowledge that can misused).

2.2. Analysis of current European regulations on ethics and participation

2.2.1. *The different institutional configurations for ethics and participation*

Merging ethics and participation brings the novelty of two fields that are not combined in the literature to the extent of the full potential of their interactions. At the intersection of ethics and participation in R&I, the process of institutionalization has led to useful categorizations that separate fields in ethics and in participatory practices. At the same time, the gray literature on the matter indicates discrepancies across European Member States.

2.2.1.1. *Ethics*

In the European R&I governance, the engagement with ethics appears in several dimensions: from ethical compliance procedures in European funding schemes to RRI frameworks. The institutionalization of ethics varies according to the level of formalization: formal engagement occurs in institutions dealing with the compliance of research integrity, or ethics in agenda-setting in science, technology and innovation; on the other side, informal engagement with ethics takes place in ad hoc activities, such as research projects (Griessler et al., 2015). Although the degree of formalization varies, this is a disputable distinction, since literature on the subject has pointed out the high degree both of precision and of development of ethical thinking in the ethics review procedure, as a formalized and thorough procedure of ethical compliance in European funding schemes.

Following the work of Peter et al. (2018), the observation of the integration of ethics across institutionalization processes sheds light on the importance of ethics committees among European Member States (half of the higher education institutions having a committee[23]), the lower influence of research integrity offices[24] and the common practice in ethics assessments of funding organizations in a number of countries[25]. Perhaps a significant

23 Except for Bulgaria.

24 They are less common in the European Union, with the exception of Germany, Belgium, the UK and to a smaller extent Estonia, Greece, Malta, Portugal and Slovenia.

25 However, this is not a common practice in countries such as Cyprus, France, Hungary, Portugal, Spain and the UK.

delineation of institutionalization levels of ethics could be considered in the types of ethical engagements, leading to a threefold distinction (Brom et al., 2015): ethical governance, ethical deliberation and ethical reflection. This tripartite distinction helps dissociate the actors and the cases that have been theoretically outlined through types of institutionalization that differ in their scope.

Connecting the various ethical approaches to innovation impact and stakeholder involvement, several directions can be outlined in the ethical analysis of emerging fields and technologies, among which two methods have been considered as particularly promising[26]:

– *ethical impact assessment*: a 14-step process beginning with determining the needs of an EIA assessment and which encompasses principles of privacy and data protection, autonomy, nonmaleficence, beneficence and justice;

– *anticipatory technology ethics*: with its three levels of ethical analysis (technology, artifact and application levels), it is geared toward emerging technologies, through forecasting and future studies for impact anticipation.

The degree and level of connection to participatory methods is however unclear in both methods, although their merits on an ethical level are clearly identified.

The institutionalization of ethics goes along with a wide array of actors who perform, fund, monitor and regulate R&I, with a growing integration of ethical considerations through ethical assessment or ethical guidance of R&I (Shelley-Egan et al., 2016). These are mainly: national ethics committees, research ethics committees, associations and networks of research ethics committees, universities and research institutes, science academies and associations of science academies, research funding organizations, academic and professional organizations in science and engineering, standardization organizations, accreditation and certification organizations, governmental organizations and councils, companies, business and industry associations, academic and professional organizations in R&I and civil society organizations (Shelley-Egan et al., 2016).

26 For more information: see the conclusions of the European projects SIENNA and SATORI.

2.2.1.2. *Participation*

At a minimal stage, public engagement could be defined as a societal commitment to provide tools to empower citizens to participate in R&I debates and processes; at a deeper level of engagement, citizens can become peers in knowledge production or assessment and government processes (Strand et al., 2015[27]). The separation between *stakeholder engagement* and *public participation* (engagement) makes it possible to distinguish the representation of specific interests (individual or collective) from the involvement of the general public in R&I activities and decision-making.

Gray literature on the subject draws the separation between these two forms that share the same aim, with different positions. The increase of public dialog regarding new developments in science and technology mainly revolves around stakeholder engagement and less on public participation, the first allowing a more granular participatory design across categories, such as end-users, citizens and representatives of interest groups. In most cases, the two categories overlap to a significant degree and are used interchangeably, even in literature that takes stock of the differences[28]. Both stakeholder engagement and public participation are expected to bring "more ethical R&I because a greater number of viewpoints and interests will be represented in them, or that ethical issues will be considered carefully" (Shelley-Egan et al., 2016, p. 32). This causal link is however questionable, as a diversity of views does not necessarily always entail more ethical results and discussion, nor a representation of diverse views in a more efficient way than with specialists.

On a European scale, public engagement is institutionalized to a varying degree across funding structures, with inequalities in infrastructures supporting engagement of citizens across countries, despite an increasing involvement of citizens in R&I processes, and also in higher education institutions (Peter et al., 2018). Furthermore, the evaluation of public engagement is considered to be absent from evaluation mechanisms in

---

27 This report reflects the work of an expert group that was appointed early 2014 by the European Commission "to identify and propose indicators and other effective means to monitor and assess the impacts of Responsible Research and Innovation (RRI) initiatives and evaluate their performance in relation to general and specific RRI objectives".

28 For instance, in the gray literature, the European project SATORI either identifies the separation of both categories or combines them without specifying: see Shelley-Egan et al. (2014, p. 16).

several countries, which indicates some limitations in its institutionalization (Peter et al., 2018). Also pointing in this direction are the variabilities of public involvement in science and technology decision-making across Europe in terms of formalization of participatory mechanisms (Peter et al., 2018). This indicator of institutionalization intersects two dimensions: the identification of formal structures and mechanisms for citizen involvement in view of decisions about science and technology; and a second dimension identifying the degree to which citizens are effectively involved in decision-making. While some countries have a formalization of participatory mechanisms and high levels of citizen participation, others have less formalized or non-formalized structures. This also shows that a higher degree of formalization entails higher involvement of citizens (Peter et al., 2018).

Other resources in the gray literature focus on the different meanings of public participation.

The European project MoRRI[29] analyzes "citizen engagement and participation of societal actors in research and innovation" (abbreviated to "public engagement"), which is conceptually defined through activities where there is a distinct role for citizens and/or societal actors in R&I processes. Public engagement is separated into five categories: public communication, public activism, public consultation, public deliberation and public participation (Mejlgaard and Ravn, 2015). The start of participation in R&I can be located in the 1960s and 1970s, when public concern with developments in science and technology rose:

> environmental depletion, consumerism, nuclear power, the dominance of multinational corporations, the risk of war etc. challenged the prevailing positive understanding of science and technology, and emphasized a need to discuss science and technology not only as instruments for solving military, economic, and social problems, but also as a source of social and environmental problems. (Mejlgaard and Ravn, 2015, pp. 10–11)

29 The Monitoring the Evolution and Benefits of Responsible Research and Innovation (MoRRI) project (2014–2018) conceptualized and implemented the first RRI monitoring system in Europe. See: http://morri-project.eu.

This led to a reconceptualization of modern science and technology "as a social activity with significant societal implications" (Mejlgaard and Ravn, 2015, p. 11), while, in parallel, the 1980s and 1990s:

> saw early, and scattered, policy responses to public concerns and scientific risks. Systematic technology assessment procedures and the establishment of dedicated organizations, ethical committees, as well as increased science communication efforts were introduced heterogeneously across countries, and in some cases, specific institutional arrangements were developed to facilitate public and stakeholder involvement in issues related to science and technology.

Correspondingly, the literature on participation has become rich and diversified. However, some general trends can be discerned:

– the field of public engagement and participation can be characterized by "a general turn from one-way and top-down models of communication towards increased focus on 'new' dialogue-based approaches" (Mejlgaard and Ravn, 2015, p. 10);

– a second characteristic[30] is that the appropriate form of participation is very context specific: where early literature on participation presented more intensive forms of participation as more desirable, literature from the late 1990s presents a more nuanced view on participation as inherently situational (Rowe and Frewer, 2000), and therefore no single best participation approach for RRI exists.

Further to categorizations proposed in the scientific literature on the subject, some recommendations for categorizations can be found on the side of the institutionalization and governance of participation. The International Association of Public Participation (IAP2, 2018) establishes a distinction for participatory practices according to their main outcomes: "inform", "consult", "involve", "collaborate" and "empower"[31]:

30 Less highlighted in the MoRRI project though.

31 See online: https://cdn.ymaws.com/www.iap2.org/resource/resmgr/pillars/Spectrum_8.5x11_Print.pdf.

– *inform* refers to governmental organizations that provide the public with balanced and objective information[32];

– *consultation* refers to obtaining feedback from stakeholders on analysis, alternatives or decisions, for example through surveys or local citizen meetings where citizens are invited to share their concerns about a certain policy;

– *involvement* means working directly with the public to ensure that the concerns and ideas of the public are considered;

– *collaboration* means that all parties collaborate as partners;

– *empowerment* refers to the process of handing over control to stakeholders by delegating decision-making power to communities (Davidson, 1998, pp. 14–15).

In addition to these recommendations for distinction within participation governance, the AA1000 Stakeholder Engagement Standard (Account Ability, 2015) aims to establish the benchmark for good-quality engagement, being a "generally applicable framework for the assessment, design, implementation and communication of quality stakeholder engagement". Claiming that its own design has been "developed using a broad-based, consultative, multi-stakeholder process", this framework is however a less useful tool for category identification, although it may be particularly informative for assessing the quality of participation.

2.2.2. *The current deadlocks of European regulatory bodies*

On a practical side, the current European landscape of regulatory bodies can reveal gaps and difficulties in bringing together ethical practices with participatory practices. A critical review of the literature and documents published in the course of the last decade provides an insight on the way different regulatory bodies and ethics bodies are dealing with participation. For this purpose, the following institutions are examined[33]:

– research ethics committees (RECs);

– research integrity offices (RIOs);

---

32 In many categorizations, this is not considered as "participation".

33 This section follows an analysis led by EUREC during the PRO-Ethics project for a study conducted by the authors of this book.

– ethics councils and ethics advisory bodies;

– the European Parliament and the European Council;

– the Council of Europe;

– professional organizations;

– NGOs and citizen organizations.

Exploring the ways that "participation" (broadly understood) is embedded in the regulations of the aforementioned bodies, a major obstacle manifests itself in the apparent overall lack of a precise definition of the participatory approach and of the objectives to which it is connected. Similarly, the vague connection to ethics does not allow for the identification of the underlying issues: neither ethical tensions, the legitimacy of participation nor the degree of formalization of participatory processes, can be clearly outlined. Indeed, the mention of "participation" in a normative or regulatory text does not necessarily provide information on the form of participation. Moreover, across the literature, a recurrent feature is the absence of clear identification of the exact addressees of the participatory mechanism. However, it can be seen that the more application-oriented a recommendation or regulatory text is, the clearer the picture of the addressees becomes, for example in the case of patient representatives. However, even this usually leaves the field open to which methodology the authors of the texts imagine for successful participation. At a general level, the literature examined indicates two broad categories of participation:

1) engagement of citizens and the public in general;

2) engagement of specific interest/stakeholder groups (e.g. patients or caretakers in the health context, NGOs or social entrepreneurs in other fields).

2.2.2.1. *Participatory practices in RECs*

Among the few examples that can be found, the resources recommend the involvement of laypersons, patients or patient organizations to reach diversity in viewpoints during research project reviews. Although participation remains undefined as a process, some categories emerge: laypersons, patients, patient organizations and the general public. As participatory approaches are not a component that is developed enough across these resources, the link between ethics and participation, as well as potential tensions, remains a blind spot of these resources; participation is identified as a general process of involvement that is needed to guarantee an ethically sound review process. The most relevant document raising this is the European regulation on clinical trials on medicinal products for human use (see European Parliament and European Council, 2014), which indicates that in the decision-making for clinical trials "at least one layperson shall participate in the assessment [of applications]" (European Parliament and European Council, 2014, article 9).

Another resource pointing in the same direction is the Guide for REC Members issued by the Council of Europe (2012), which underlines that:

> The appointment mechanism should ensure that potential REC members provide an appropriate balance of scientific expertise, philosophical, legal or ethical backgrounds, and lay views. All REC members, whether professional or lay members, should have an equal standing. (Council of Europe, 2012, p. 19)

2.2.2.2. *Participatory practices in the field of research integrity*

Several research integrity bodies and representative institutions of this field in Europe cover ethical regulations. These are, among others: ENRIO; the World Conference on Research Integrity (WCRI)[34]; the European Network for Academic Integrity (ENAI)[35]; the European Network for Ombuds in Higher Education (ENOHE)[36]; the UK Research Integrity Office (UKRIO)[37]; the Austrian Agency for Research Integrity (OeAWI)[38]; and the Czech Academy of Sciences[39]. Considering the work undertaken by these bodies, their resources (codes, notably) underline the importance of ethical principles (fairness, transparency, gender, privacy, sustainability), which are the basis for citizens' engagement. However, none of these bodies has drawn a meaningful connection with participatory practices in a comprehensive manner.

---

34 See online: https://wcrif.org.

35 See online: https://www.academicintegrity.eu/wp/.

36 See online: http://www.enohe.net.

37 See online: https://ukrio.org.

38 See online: https://oeawi.at/en/.

39 See online: https://www.avcr.cz/en/.

2.2.2.3. *Participatory practices in statements from ethics advisory bodies*

The most relevant statement on participatory practices is the one published by the European Group on Ethics in Science and New Technologies (2015) *on the ethical implications of new health technologies and citizen participation*. This publication focuses on citizen science perceived as an "active participation" form, which is an interesting distinction of participatory quality[40]. Also, it recognizes and encompasses several individual or organized actors: "stakeholders, lay persons, patients and consumers but also and in counterpoint, organised interest groups, lobbies and corporate bodies" (European Group on Ethics in Science and New Technologies, 2015, p. 24). Other resources have shown variations in the identification of participation and participants, depending on the relevant field of application: while the health sector targets patients and citizens, the sector of big data, for instance, usually refers to the involvement of the general public.

2.2.2.4. *Participatory practices in codes/guidelines on emerging technologies*

A sectoral analysis in the field of artificial intelligence (AI) and robotics (codes and guidelines) shows a general approach to participation, encompassing the relevant public authorities and stakeholders, society as a whole and local communities. In these two emerging technology sectors, participation is mostly perceived through the spectrum of future impacts on society: the people (or groups) who might be affected by developments of AI and robotic technologies. Three main documents support these assumptions:

1) the IEEE Code of Ethics (2006)[41] mentions the need to provide adequate public information so that individuals and society may understand the potential impact of emerging technologies;

2) also emphasizing public information, the Montreal Declaration for a Responsible Development of Artificial Intelligence (University of Montreal, 2017)[42] identifies key principles, among which a "democratic participation

---

40 We will come back to the quality of participation at the end of this chapter.

41 See Section 7. Retrieved from: https://www.ieee.org/about/corporate/governance/p7-8.html.

42 Retrieved from: https://www.montrealdeclaration-responsibleai.com.

principle" mentioning that artificial intelligence systems "must be subjected to democratic scrutiny, debate and control";

3) another statement on the importance of community engagement can be found in the Humanitarian UAV Code of Conduct[43], which mentions trust building with local communities as a pillar, which also allows communities to be active participants[44].

Overall, the study of resources across ethics regulatory bodies reveals a rather weak connection between ethical practices and participatory practices: the connection is often unspecified, maintained at the level of a general appreciation of potential benefits[45]. In the absence of precisions on the participatory practices themselves, the mention of public involvement generally refers to public information initiatives, without further elaboration. Also, participants themselves are rarely distinguished, being globally identified either as "citizens", "lay people" or "patients" (in the health sector). However, two exceptions to this are: first, the mention of citizen science, which refers to a specific form of public involvement (whose connection with decision-making has nevertheless to be clarified, in each case); and second the mention of research participants which also refers to a specific form of participation where citizens are also themselves part of the R&I process. Considering the different fields of R&I, the sectoral specificities of the health sector and of emerging technologies (AI and robotics) evolve in the direction of public information for potential/future impacts of R&I processes and products on society[46], proving the advancement of these fields in the inclusion of participatory practices as a pre-requisite in most cases.

---

43 This code of conduct informs the responsible use of civilian drones or unmanned aerial vehicles (UAVs) in humanitarian settings: it bridges emerging technologies and humanitarian applications. See Humanitarian UAV Network (2015).

44 This code of conduct mentions that "building trust with local communities allows them to be active partners, decision-makers and enablers, thus enhancing the mission and humanitarian/development impact".

45 Usually justified by the wish to keep a transparent relationship with society across R&I processes.

46 Here, too, the connection with decision-making varies.

2.2.3. *Ethics of participation: current boundaries*

On the side of European policy-making, the main limits in the ethics of participation[47] are to be found in the operationality of ethical frameworks. Participatory governance ensures stakeholders are able to participate in solving ethical issues that shape the future of our societies, as public engagement fosters social commitments and strong social outcomes. If categorizations of participation are a useful tool that can help sharpen the efficiency of policy-making in this field, however, "a further step involves understanding and defining, perhaps via a second typology, the different types of *context* in which engagement takes place" (Rowe and Frewer, 2005, p. 286). The categorization entails that the selection of a participation mechanism derives from a clear identification of the context and the anticipated effects: contextual criteria have to be measured in the final visible effects of a decision-making process (Shelley-Egan et al., 2014).

The ethical governance of R&I appears as a long-shot ideal as long as norms and paradigms are confronted with multifaceted approaches and contextual criteria that do not allow for a monolithic analysis. In addition to this, the uneven level of ethical or technical expertise implies a different response to ethical issues encountered in R&I developments (Kurt et al., 2010), while the predominant views of technical experts in governance may hinder the development of ethical issues in a participatory way. Also, the reduction of ethics to consensus or to regulatory framings is another obstacle, as moral values are often "compromised for the sake of consensus in reaching an ethical norm" (Kurt et al., 2010, pp. 45–46). As a consequence of this, stakeholders' involvement in the ethical governance of projects might be reduced to poorer forms of participation, with the stakeholders' role being limited to feedback mechanisms to inform the design process instead of "addressing and solving the ethical issues, which would mean adjusting the technology or in extreme cases abandoning it" (Kurt et al., 2010, p. 46).

Seeking enhanced participation that goes beyond unspecified recommendations by regulatory bodies or the ethics review formalized scheme – limited to R&I funded by the EU – the path forward seems to be dependent on the level of clarity regarding the context and the participatory features considered. However, even well-advanced mechanisms such as

47 Referring here to a broad definition of participation.

ethics reviews are confronted with a number of challenges reported by RECs and RFOs[48]:

– a *lack of clear procedures* (standards, protocols, guidelines, tolls) for ethics assessment;

– *heterogeneity in approaches* and guideline implementation;

– *overloaded ethics committees*, lack of fruitful discussion, human factors, inconsistent reviews;

– *lack of resources* (financial, human, time, knowledge)[49].

An additional limitation in policy-making is the timeframe of the development of R&I projects. These latter are intrinsically confronted with the timing gap with legal and ethical compliance, as the second does not necessarily anticipate all developments, hence the need to adapt the regulatory environment for innovation, making sure that the timeframe of legislative processes can follow the fast-evolving landscape of technologies, as well as the difficult combination of innovation-friendly regulations with the precautionary principle (European Commission, 2016).

Gray literature resources also focus on good practices in participatory processes as part of ethics-related projects in order to anticipate some difficulties. This can be handled by covering all stages: the preparation, the design, the implementation and the follow-up of participatory processes[50]. A Proper preparation entails the clear identification of the goals, as well as a wise selection of stakeholders, target groups and the participatory approach, so as to reach a process that aims toward a co-construction in strategy-setting and decision-making (Shelley-Egan et al., 2014). On the design level, a bottom-up and flexible approach allows engagement with stakeholders on the process and the results so as to reach representativeness and generate added value (Shelley-Egan et al., 2014). During implementation, having a good facilitator is an essential element that, however, should not undermine the contribution of lay people, as the main issue is to have a balanced and

48 It might be important to note here that in research funding organizations, ethics assessment of research proposals is not always formalized.

49 The full list and a comparative approach can be found in Shelley-Egan et al. (2016, pp. 80–82).

50 See the Stakeholders Acting Together on the Ethical Impact Assessment of Research and Innovation (SATORI) project (2014–2017) aimed at developing a common European framework for the ethical assessment of R&I. See online: https://satoriproject.eu.

open process. The follow-up phase (after the participatory process) should provide an analysis of the outcomes of the process and receive further feedback. However, these recommendations do not outweigh the potential technical difficulties that may arise, especially the ability to identify the degree to which participatory activities can have a direct influence on the policy-making process (Shelley-Egan et al., 2014, p. 8). Also, the varying levels of ethical engagement and the lack of a comprehensive approach to participatory mechanisms across R&I processes (which differ in nature and timing: before, during or after R&I processes) cannot be easily solved. As already analyzed (Shelley-Egan et al., 2014), European policies are supported by strong normative assumptions: in R&I, policies rely on weak normative assumptions, and with the ongoing tensions with ethics, are often seen as a constraint. The way toward unified European policies with strong normative assumptions encompassing innovation is still a worksite.

2.3. Soft law on participatory approaches

2.3.1. *Re-shaping participatory approaches through soft law: toward a normative instrument*

2.3.1.1. *A normative basis*

On the side of existing regulations, the main reference on participation is the 1998 Aarhus Convention on Access to Information, Public Participation in Decision-Making and Access to Justice in Environmental Matters (United Nations Economic Commission for Europe, 1998), which establishes the rights of the public with regard to the environment. Despite the focus on environmental matters in this Convention, the general principles could arguably serve our general considerations on regulatory frameworks concerning participation. The sixth article of this Convention could be considered as pivotal, since it specifies the various kinds of public participation in decision-making through the following distribution:

– public information[51];

– public involvement procedures[52];

– integration of public participation outcomes[53].

---

51 See Article 6 §1; §2; §9.

52 Article 6 §3–7.

53 Article 6 §8.

These aspects are complemented by some specificities further to public involvement procedures, notably in the process of program and policy preparation[54] and the preparation of legally binding normative instruments[55].

The Aarhus Convention might be the only normative instrument of its kind addressing the issue of participation directly, which, albeit the specific sectoral focus on environmental concerns, is outlined in a clear breakdown of its characteristics and implications. Given the operationality of this instrument and the clarity it brings in the distinction of participatory processes, it might be considered as quite relevant for other sectors in R&I. Indeed, the uptake of recommendations on participation revolves around the understanding and consideration of the tripartite distinction of participation as information, consultation and participation in decision-making. This process of methodical partition between participation processes can be taken as a reference and extended to innovation processes across all sectors.

2.3.1.2. *Policies and soft law*

On the level of policies, citizen science frameworks range beyond the scope of normative instruments and yet can be valuable in defining governance tendencies in participatory practices as a well-defined area. The reign of "do-it-yourself scientists" (Nascimento et al., 2018, p. 236) doing science outside conventional spaces could be included in ground-breaking features shaping future participatory modes, based on bottom-up approaches and co-creation through citizen empowerment and inclusion. This reflects the wider tendency of the science–society relationship, which has shifted from top-down traditional approaches (expert governance) toward bottom-up models of public engagement anchored in dialog and the co-construction of R&I agendas, bringing new institutional models and a democratic turn, thanks to the new role of the public. Citizen science seems to contribute to this tendency, although we may consider that it mostly represents institutionally led (top-down) initiatives (Nascimento et al., 2018).

The current great interest in participatory mechanisms engaging citizens in science and technology raises questions on the ways to encourage meaningful engagement in top-down configurations and on how to build independent and effective citizen engagement capacities. While a number of

54 Article 7.

55 Article 8.

initiatives have sprung up over the last decade, existing engagement models suffer from limitations, the majority of them being unable to efficiently provide lay citizens with the skills and capacities to engage and have an impact in scientific and technological developments (Powell and Colin, 2009). Depending on the goal, citizen engagement can be meaningful – for example to reach public acceptance or to reach policy impact – but is not a goal as such (Powell and Colin, 2009). Following this, bottom-up active citizen engagement may not be desirable or meaningful, considering, for instance, power differentials between lay citizens and scientists[56]. The engagement of lay citizens finds two cases of justification: (1) they should have a say in scientific and technological development that will affect their lives and the broader society in significant ways; and (2) they bring valuable knowledge and perspectives into decisions and this diversity makes decisions more robust (Powell and Colin, 2009).

The institutionalization of citizen science has led up to the conception of the Ten Principles of Citizen Science, a framework developed in 2015 by European Citizen Science Association (ECSA), an international community of citizen science practitioners and researchers setting out characteristics of high-quality citizen science, highlighting good practice (Robinson et al., 2018) to both support and challenge the citizen science community and to improve practice. The core principles promoted are based on openness, accessibility, meaningful participation and recognition of contributions (Robinson et al., 2018), and the involvement of citizens can occur in multiple stages of the scientific process. Although incorporating local knowledge and expertise can benefit the research, the impacts of co-created citizen science have not been assessed enough (Robinson et al., 2018).

Citizen science acts as a bridge between citizens and policymakers, it is renowned as a useful source of information for governments and it has been proven effective in fostering open science, thus contributing to policy design and implementation. However, the mechanisms for citizens to impact evidence-based processes for policy-making are still lacking (Nascimento et al., 2018). On a general level, several questions arise on the topic of citizen science regarding the different approaches (contributory, collaborative, co-created) and

56 According to Maria Powell and Mathilde Colin's investigation with a citizens' coalition on nanotechnology, citizen engagement can be initiated from the top-down but requires significant top-down guidance, which is challenging and time consuming. See Powell and Colin (2009, pp. 327, 331).

the implementation of such projects on a large scale, while maintaining interaction with participants throughout the scientific process (Robinson et al., 2018).

The potential citizen science contributions to policy are mainly as follows:

– meeting the data collection targets of programs that need to monitor large geographical areas with high frequency;

– providing evidence for assessments by supporting regulatory compliance;

– community empowerment and awareness raising (Nascimento et al., 2018).

These benefits are contributing to the general consideration of citizen science as a timely, cost-effective source of information aiming at supporting evidence-based policy implementation (and monitoring) and complementing official sources (Nascimento et al., 2018). Indeed, the growing interest in citizen science and its proven economic value, thanks to in-kind contributions, favor its further development within policies. Furthermore, it is recognized as instrumental in fostering novel science in R&I strategies, policies and initiatives and ensures R&I agendas are guided toward issues of concern to citizens (Nascimento et al., 2018). However, some negative aspects of citizen science can be identified, such as the fact that it does not necessarily entail more participation of citizens and does not ensure the participation of local communities. Nevertheless, it acts as a powerful citizen mobilization and can bring a change in attitudes, since the involvement of public and civil society stakeholders in co-creation fosters public acceptance (Nascimento et al., 2018).

2.4. Beyond regulations: institutional needs for ethical participation

2.4.1. *The needs from a governance point of view*

Among the great variety and number of public engagement mechanisms[57], the identification of best practices is not easy as there is uncertainty "as to how one should *best* enact involvement" (Rowe and

57 Views diverge on the number/categories of techniques. See Rowe and Frewer (2005, pp. 251–290).

Frewer, 2005, p. 252). Only a theory of the contingent effectiveness of engagement mechanisms can be developed, in light of the inherent variability, as one mechanism is unlikely to be the most appropriate/effective in all situations (Rowe and Frewer, 2005). The effectiveness of public engagement depends on the mechanism and the way it is applied, along with the presence of *active*, as opposed to passive participants (whose opinion is not heard during the participatory process), as well as the aggregation process and its efficiency – since the equity in input is not guaranteed in group-based output where, according to the procedure, polarization and influence can occur diversely (Rowe and Frewer, 2005).

Through formalized forms of public engagement, feedback on effective experiences with metrics of efficiency allows concrete evidence for policy-making or what could be referred to as "soft governance". In the absence of precise indicators, such metrics can rely on criteria which help determine the pros and cons of main current participatory processes of public engagement. Seeking democratic choice and consent, public participation should be framed through precise criteria that can ensure both its legitimacy and efficiency. Regarding the issue of acceptance, the following criteria could be considered (Rowe and Frewer, 2000):

– *representativeness*: participants should comprise a broadly representative sample of the population of the affected public and represent the relative distribution of views;

– *early involvement*: the public should be involved as early as possible in the process as soon as value judgments become salient;

– *influence*: the output of the procedure should have a genuine impact on policy;

– *transparency*: the process should be transparent so that the public can see what is going on and how decisions are being made.

In terms of process, efficient public participation could align on the following criteria (Rowe and Frewer, 2000):

– *accessibility*: public participants should have access to the appropriate resources to enable them to successfully fulfill their brief;

– *task definition*: the nature and scope of the participatory task should be clearly defined at the outset, so that there is as little confusion and dispute as possible regarding the scope of a participation exercise, its expected output and the mechanisms of the procedure;

– *structured decision-making*: the participation exercise should use/provide appropriate mechanisms and tools for structuring and displaying the decision-making process;

– *cost-effectiveness*: the concern of cost in participation methods should be addressed, so that the procedure can be cost-effective.

Formalized participation models respond differently to these different criteria, and while some of them have low rates in terms of acceptance and process criteria such as the public hearing model, others respond well to one category: referenda, public opinion surveys and focus groups respond well to acceptance criteria but not to process criteria. On the contrary, other participatory approaches such as consensus conferences, citizens' juries/panels and citizen advisory committees meet both process and acceptance criteria (Rowe and Frewer, 2000).

Expanding the dialog with the public – from an optional add-on to an integral part of the process of policy-making – requires re-designing the democratic process in a more balanced way, where constituents are also active participants (Fung, 2008). At the government level, participation takes the form of "public engagement" and has gained great interest over the years, through various methods aiming at a culture of openness to ideas through various forms: large-scale government-led exercises, platforms for engaging citizens, methods for participation, idea generation and deliberation, processes for involving children, "wiki government", participatory planning, parliamentary structures to develop citizen ideas, citizen petitions, citizen juries, citizen's panels and legislative theatre (Murray et al., 2010).

The identification of the various participatory approaches, the timescale of intervention and the objective, whether it concerns evaluation, planning or implementation, can help dissociate active forms of participation from other indirect forms and thus reach a concrete framing of the process. The choice of the participatory method can rely on a set of criteria relating to objectives, topic, participants, time and budget, which entails considering: (1) the reasons for involvement and the expected outcomes; (2) the nature and scope of the issue; (3) who is affected, interested or can contribute to solutions; (4) the amount of available time; and (5) the availability of resources (Slocum, 2003).

Reaching clarity in the definition of the objectives is also a key element in the efficiency of participatory processes, ensuring their adequacy with the expected outcomes. On a general level, Slocum (2003) identifies two main categories: an "aspiration level" aiming at a democratization process or an advising process[58]; and an "output level" aiming at a diversification of options or consensus-reaching.

The identification of participants is also essential in the selection of the participatory process as categories have to be clearly distinguished. Slocum (2003, p. 13) identifies the following categories of participants:

– *citizens* (as individuals);

– *stakeholders* (citizens represented by organizations, such as: NGOs, private industry and interest groups);

– *experts* (for the issue considered);

– *politicians* (for the uptake of the outcomes);

– *policymakers* (this category is particularly relevant for processes that are likely to influence policy).

On the timeframe level, and similarly to the distinction of timescale in technology assessment, in policy issues too participation differs significantly in terms of impact, according to the moment at which it takes place. In "active" forms of participation, looking for feedback loops from end-users and citizens can help better address societal impacts from the stage of initial design and planning (Slocum, 2003). While achieving consensus can be seen as a pitfall in trying to exert inclusive and active participation, early engagement can allow equal sharing of perspectives, values and reasoning (Slocum, 2003). An additional criterion helping in reaching adequacy between the issue and the way it is addressed is defining the need for a participatory approach, which relates to the following cases (Slocum, 2003):

– themes that require ethical, social or cultural study and may call for a choice between fundamental values and principles;

58 According to Slocum, democratization aims at enabling participants with information to make an active contribution by finding solutions relevant for decision-making, while advising aims at revealing stakeholders' knowledge and positions relevant to the process of decision-making.

– policy issues that call for a combination of public awareness, learning, a search for solutions and emotional or moral acceptance of the eventual decision;

– public policy choices that rely on the precautionary principle or the weight of evidence;

– underlying values and principles that must be clarified before detailed proposals or risk management options are brought forward;

– a clearly defined set of options or proposals that support the search for consensus or innovative solutions.

2.4.2. *Identifying good practices and criteria for ethical participation*

If "ethics should not be perceived as a constraint to R&I, but rather as a way of ensuring high quality results" (Strand et al., 2015, p. 33), the participatory turn can help sustain such views, as public participation entails that "upon everyone's shoulders rests a particular moral obligation to engage in the collective debate that shapes the context for collective decision-making" (Von Schomberg, 2007, p. 12). Governance and funding maintain strong leverage in shaping innovation practices toward an ethical and participatory turn that takes stock of context specificities to define adequate approaches. A responsible innovation strategy should intervene at all stages of R&I, encompass all ethical issues and ensure a continuous inclusion of stakeholders' values in the process of technological design, leading to a process of extraction of relevant public values and their input into technological design, as well as the work of relevant institutions and decision-making processes (Taebi et al., 2014)[59].

An upstream connection of science and society during R&I processes allows for a connection of ethical issues with participatory approaches. The ethics of collective co-responsibility, "expressed at the level of free (international) public debate in which all should participate" (Von Schomberg, 2007, p. 11), can be addressed in such early engagement configurations. Early engagement can be taken a step further by including

59 However, the implementation of such a process faces difficulties stemming from moral pluralism implications.

both people with a different (social science or humanities) background in the innovation process as well as lay people/civil society.

Following the democratic–inclusive paradigm, which combines democratic participation of a community of citizens and the inclusion of society in the determination of social options (Kurt et al., 2010), the democratic governance model entails – further to consultation – the co-construction of rules and options that matter for society[60]. Related to the democratic paradigm, the co-construction model of governance involves stakeholders and policymakers in the construction of policies, while the role of experts is extended to more participants in a participative approach to innovation[61].

The way forward requires a combination of an ethical framework addressing "both the aspect of unintentional side consequences (rather than intentional actions) and the aspect of collective decisions (rather than individual decisions) with regard to complex societal systems"; the lack of such theoretical guidance inclines us to "shift our attention to an ethics of knowledge assessment in the framework of deliberative procedures instead" (Von Schomberg, 2007, p. 5). Essential components of responsible innovation frameworks are the result of a strong combination of ethical concerns with an active use of participatory challenges:

– interdisciplinary research;

– public values: ex ante assessment of stakeholder values and the specification of values during development and implementation[62];

– challenges to give weight to varying opinions;

– facilitation of trade-offs (timely and proactive identification of potentially conflicting values) (Taebi et al., 2014).

Another guidance tool can be found in the potential of social innovation and governance structures that foster sustainability (Howaldt and Schwarz, 2010) and aim at "finding better and alternative ways to meet existing needs

60 Although such approaches can be confronted with difficulties in bridging the knowledge gap between experts and lay persons on technical topics.

61 On the topic of the "co-construction model" of governance; see Kurt et al. (2010, p. 40).

62 According to these authors, other approaches such as participatory technology assessment are not equipped to deal with the values that emerge during technology development and implementation (Taebi et al., 2014, p. 121).

and to more effectively work through unintentional repercussions and side effects of industrial development in society" (Howaldt and Schwarz, 2010, p. 45). RRI provides a substantial basis to build on, in that by integrating its principles into policy, participation can be leveraged, with a focus on the dimension of responsiveness.

Deliberation opens up the degree of uncertainty and reveals the risk-taking dimension of participatory approaches. As future developments and their outcomes are both scientifically uncertain, they become increasingly indeterminate with participation, "as they are shaped reflexively through an open horizon of contingent choice and political debate" (Demeritt et al., 2011). Furthermore, the difficulty of establishing balanced ethical frameworks has already been underlined in the literature as the difficulty of choosing between strong and weak definitions of participation or "between anthropocentric and ecocentric moral frameworks for pathway evaluation" (Demeritt et al., 2011).

On the side of funding bodies and governments, the notable improvements in the "opening up" of their governance processes are often still constrained by the fact that:

> dialogues still tend to be seen as an add-on to established structures, rather than the start of a new sort of relationship with the public. There is, therefore, a need to move beyond thinking of public engagement in isolation, to talk about governance in the public interest. (Sykes and Macnaghten, 2013, p. 104)

Contradicting the tendency in science study to resolve issues of governance through normative treatment (Irwin, 2006), participation shows another way in open science–public relations that can bring significant legitimacy to R&I processes.

The shift from R&I processes *from* the public to R&I processes *by* the public has marked a new participatory paradigm that goes beyond mere consultation or information mechanisms toward collaborative and transformative mechanisms where the public takes part in decision-making and is an active part of the process. In RRI, the focus on inclusive deliberation pledges the involvement of a diverse range of engaged stakeholders and publics, so as to increase legitimate decision-making in regard to "socially robust knowledge" (Owen and Pansera, 2019, pp. 31–32). The related goals and outcomes bring about, among other results, the opportunity of participation in agenda-setting and defining societal challenges; equitable decision-making; and a better capacity and basis for robust and legitimate decision-making (Owen and Pansera, 2019). Through the prism of a shared co-responsible approach of innovation, deliberative systems are facing the constraint to serve a wide array of values, as they could:

> be judged according to how well they serve the epistemic, ethical, and democratic functions of deliberation. Epistemically, they should produce preferences, opinions, and decisions that are appropriately informed by facts and logic and that derive from the meaningful consideration of relevant reasons. Ethically, deliberative systems should produce mutual respect among citizens. Democratically, deliberative systems should give voice to multiple and plural perspectives, interests, concerns, and claims on the basis of feasible equality and equal opportunity. From all three perspectives, a healthy deliberative system is one in which relevant considerations are brought forth from all corners, aired, discussed, and appropriately weighed. (Mansbridge, 2014, p. 240)[63]

A comprehensive deliberative form would require a complex setup in the coordination and framing, in order to ensure a diversified landscape with a fair distribution in the visibility of perspectives and views. If we may consider it an ideal configuration, its implementation and operationality would also entail the reinforcement of information sharing among participants, as well as the expansion of space for a diversified debate in views and in sharing capacities and with an agreed common respect of ethical principles, norms and values. Despite the technical difficulties of this ideal, it provides a sound basis of directions that could be considered as general orientations for a pluralistic participation.

At the same time, new perspectives are emerging with the expansion of digital technologies, which have propelled online platforms using participatory approaches for new costless solutions (enhancing the relationship between citizens and governance), allowing for citizen–science

63 It could be argued that a threefold compliance of deliberative mechanisms with epistemic, ethical and democratic values could be considered either this way or perhaps more appropriately as a whole, since all dimensions overlap as ethics, which also embraces the subjects at stake and their conceptual representation in the debate.

collaborations, crowdsourcing and online surveys (citizens as subjects of R&I processes)[64]. Although so-called "civic tech" has emerged as a promising new field for participatory practices, the variabilities of civic tech movements, in terms of technology used, degree of change, publics engaged, social processes and functions, do not allow for a unique definition[65]. The growing and diverse field of civic tech has emerged at the nexus of technology, civic innovation, open government and resident engagement (Patel et al., 2013)[66]. The great variety of initiatives termed "civic tech" hinders the direct connection with RRI practices, although the two main forms of open government and community action[67] bear in themselves the main features of participatory practices. If we consider the specific case of R&I, the first feature, which is reflected through citizen consultation and data transparency on behalf of governments, reinforces the democratic governance of innovation. The contribution of citizen-led initiatives on the other hand can considerably improve the conditions for citizen empowerment and collective decision-making.

Civic tech innovation can be identified in 11 main forms (Patel et al., 2013), across the two dimensions. On the side of open government:

1) *data access and transparency*: promotion of government data availability, transparency and accountability;

2) *data utility*: empowerment of users to analyze government data and leverage data to improve public service delivery;

3) *public decision-making*: encouraging resident participation in large-scale deliberative democracy[68] and community planning efforts;

4) *resident feedback*: providing residents with opportunities to interact with government officials and give feedback about public service delivery;

64 Some principle online platforms and projects worldwide that represent this new trend of online participative science and innovation are listed in the French report by Houllier and Merilhou-Goudard (2016).

65 Microsoft's innovation team has produced the Civic Graph, a crowdsourced interactive online map tracking the new world of civic tech (civic innovation). See online: https://www.civicgraph.io.

66 This publication issued by the Knight Foundation relies on large-scale mapping of the civic tech field by combining semantic analysis and investment data tracking.

67 The Knight Foundation's study identifies them as overarching themes.

68 The tools to achieve this represent a major challenge, although the evolution of civic tech features seems to be promising in this regard.

5) *visualization and mapping*: enabling users to make sense of and gain actionable insight from civic data sources, specifically through the visualization and mapping of that information;

6) *voting*: supporting voter participation and fair election processes.

On the side of community action:

7) *civic crowdfunding*: supporting local projects and organizations that generate a public benefit through peer-to-peer lending and crowdfunding;

8) *community organizing*: managing social campaigns and initiatives;

9) *information crowdsourcing*: collecting data from a large number of individuals to inform and address civic issues;

10) *neighborhood forums*: empowering local groups of people to connect, share information and collaborate;

11) *peer-to-peer sharing*: promoting resident-driven sharing of goods and services (Patel et al., 2013).

The predominance of peer-to-peer sharing initiatives in the global growth of civic tech hinders the variety of the field and the assets it offers in terms of the embeddedness of deliberative democracy features with data access, as well as diversity and the costless access to a toolbox for participants' inclusion.

The connection with RRI attempted here aims to initiate a new reflection on the leverage of civic tech projects in the promotion of participatory practices for RRI.

2.5. Conclusion

From the viewpoint of binding and non-binding legal frameworks, the analysis of the way regulatory bodies across Europe deal with participation underlined the difficulties and limits policy-making is facing. The regulatory frameworks pertaining to ethics of innovation are mostly related to research ethics and research integrity. Inasmuch as research participants are involved, participatory practices may find a substantial basis on that front.

Besides main regulations, formalized ethics assessment in European funding schemes provides interesting insight on how ethical features can effectively frame R&I processes in a case-by-case basis. The current regulatory landscape in Europe embraces participation in a rather undefined way, usually alluding to participatory features without making the distinction between participant types and the way to ethically conduct such initiatives. Similarly, while some sectors such as health and emerging technologies benefit from more clarity regarding participatory practices, the overall unspecified connection between ethics and participation is a general trend across regulations. Past studies regarding the limitations of ethics procedures at the European level indicate a poor development of structural means (lack of resources), a lack of guidelines and weak capacities on behalf of ethics committees to cope with this situation.

Acknowledging these current limitations in the field of ethics regulations and procedures sheds light on the weak connection between ethics and participatory practices. This suggests the need for an extension toward soft law and guidelines on participatory practices that can be found in the scientific literature. The challenges that have been presented so far can be faced through criteria and good practices that can indicate directions for future frameworks and new methodological pillars serving to guide participatory practices in an ethical manner.

3

Ethical Participation

The combination of ethics requirements with the complexity of participatory practices appeals to the reinforcement of existing institutionalization processes in the direction of classifications and indicators. In line with the original focus of this study, namely, the governance of R&I, this last chapter focuses on the ways to enable institutionalized participatory practices with the tools allowing for an ethical approach to participation. Among all R&I institutions, at local, national or European levels, funding organizations are an interesting focus in that, regardless of their operational mechanism, they can use funding as a lever for the acceptance of ethics frameworks. This chapter draws conclusions retrieved from a case study combining first-hand data from European RFOs at European levels.

A core assumption here is that to a certain degree integrating ethics features and guidance as part of a framework adopted by R&I institutions can only benefit the current regulatory vacuum. This latter entails either unfitted participatory processes (in terms of legitimacy, and/or outcomes, and/or connection to decision-making) or difficulties during the implementation of these processes. Such guidance can only take the form of an ethics framework, whose application can be left to the institutions' discretion. Within the wide range of R&I institutions, RFOs are the focus of this chapter, thanks to the very specific insight they provide on formalized ethics procedures, as well as regarding the use of funding as a leverage in ethical endeavors. From then on, ethical features related to participatory practices can be perceived as complementary to the existing ethics procedures. Such a framework is considered as overall guidance, stemming from the actual needs and expectations of RFOs.

As potentially great levers, RFOs in Europe have specific needs and expectations in regard to participation, which reveal the added value that participatory practices can have in R&I processes as closed expert-led circuits (section 3.1). Indeed, although benefitting from public attention, citizen participation is not necessarily prevailing in institutional processes, as shown through a case study presented below. The identification of different participatory activities and participants allows for a precise identification of some prevailing forms in R&I processes in European RFOs. R&I institutions have to deal with blind spots when designing or implementing participatory practices; however, the main obstacles remain their structural capacity to *ethically* design or implement such practices (section 3.2). Responding to criteria of legitimacy and impact goes along with the expectations from the organizers of participatory initiatives, with a varying connection to ethics. Working toward an ethical basis (section 3.3) then implies considering the current gaps and combining them with new methodological features that can help address each specific R&I process with a stable analysis grid to guide participatory practices. In line with the analysis of institutional needs in the previous chapter, we here take stock of the needs of RFOs, and their views on the gaps effectively enabling ethical participation. Further to this, some key factors are presented (section 3.4) as main pillars worth considering in any future institutional endeavor toward an ethical approach to participation. This is complemented by the proposal of a tentative ethics framework, as part of a real-life testing pilot[1] (section 3.5).

3.1. How mainstream is citizen participation?

3.1.1. *A case study on research funding organizations*

An analysis of RFOs' practices in Europe was conducted under the PRO-Ethics project[2] through a survey with additional interviews to further analyze the governance structure of RFOs, their mechanisms, their views on ethics and on participation, their capacities and aspirations regarding future ethical development and the development of participatory practices and citizen engagement in particular. This study brought forward interesting

1 In the European project PRO-Ethics.

2 European project supported by the European Union's Horizon 2020 research and innovation program, running from 2020 to 2023. More information is available on the project's website: https://pro-ethics.eu.

considerations on the ethical practices and the features of participatory processes in public funding schemes for R&I in Europe. Overall, the data from the survey have shown prevailing concerns in legal compliance, as well as expert-led governance of R&I across the main RFO activities. The limited difference between current practices and future expectations indicates a narrow capacity for change, both in the contribution of ethics and in the way participatory processes are designed. The interviews with RFOs[3] allowed for a close-up view on their aspirations and the constraints they are facing in their future development.

A sample of 10 organizations was selected, drawing from the PRO-Ethics consortium's RFOs[4]: the survey was designed with an analysis grid that was built for this very purpose, as a first step toward future taxonomies. The choice of the consortium's RFOs was motivated by the fact that they could effectively provide qualitative data and detailed background analysis, which were targeted as core and indispensable elements in the methodological framing prior to the launch of this study[5]. The overall aim of this report as well as the other empirical component (interviews with experts) has remained unchanged. The results and the quality of the analysis in this study benefitted from the aforementioned adaptation in the selection of RFOs. The detailed survey questions were developed thanks to this adapted selection, aiming at maximizing the overall goals of the empirical analysis in terms of findings. The survey results reached a high degree of detail, which had a direct impact on the qualitative and quantitative assessment of the different features considered in this study regarding RFOs' practices.

Among these organizations were eight RFOs and two organizations at the peripheral level – an international organization and an expert organization –

3 Conducted in the course of the PRO-Ethics project.

4 Instead of a large-scale quantitative survey with European RFOs (as a target group), this case study focused on a smaller qualitative approach with collaborating RFOs as target group in order to retrieve more data and conduct an in-depth analysis of their practices.

5 This batch of RFOs had been already selected at the proposal stage in PRO-Ethics for its structural and geographical diversity, as well as for the commitment of each RFO to the project, which could easily secure their position as active contributors with a stake in the quality of the mapping.

neither of which are labeled an RFO, but both with substantial funding activities. These RFOs deal differently with program and sectoral priorities, regulations and ethical considerations, and they interact differently with civil society, private sector actors, academic actors and various other stakeholders. Their understanding of "participation" varies to a certain extent in terms of perimeter of action, needs and practices, while their own definition and identification of participants is as diverse as the ethical grounds of their practices. The survey combines quantitative with qualitative, more open-ended items, as well as interviews with each RFOs' leading team, aiming at collecting and comparing the practices and expectations, the legal and ethical grounds that are part of the operational model and the national specificities that are faced by the RFOs. The first version of the survey, cognitively pre-tested, was repeatedly amended and re-shaped in its entries and layout in order to eventually obtain the qualitative data that were expected for this study.

Through the survey, an initial mapping of the diversity of definitions of ethical practices and participatory processes highlighted the discrepancies, both in current practices and future expectations, as well as certain trends in terms of main categories, uses and needs in participatory approaches in R&I. Also, this survey helped to draw a categorization of modes of participation and outline the perimeter of what is meant by "participatory approaches" in R&I activities.

Some ethical challenges or issues related to participatory approaches that are critical to RFOs were examined more closely in the course of the RFO interviews following the survey: identification of participants, representation of participants, identification of the right process for participants' involvement, adaptation of participative methods to the context, avoidance of biases in citizen selection, use of personal data, consideration of the asymmetric access to information for citizens (awareness of participation) and also the fact that not all RFOs consider ethics as a meaningful component.

In the RFO survey, contextual variabilities had to be addressed relating to top-down approaches and the varying flexibility of approaches, the nature of funding programs and the kind of community each RFO is dealing with. What are the main challenges and benefits of engaging citizens or stakeholders? What are the most important aspects when undertaking participation? How should participants be selected? Such questions guided the empirical data collection, as well as the criteria selection and the methodology used to display the results of this study.

For the RFO survey, data were collected through a survey combining quantitative and qualitative questions and both open and closed questions, complemented by interviews. The categories of the survey aimed at a granular approach to participation actors and modes. These data were further developed with qualitative questions, serving for the background analysis. The set of questions on current practices was symmetrically deployed for future expectations, in order to point out major trends identified by RFOs in their institutional development. In addition to these data, individual interviews with each RFO provided some more contextual elements that helped to point out national specificities as well as specificities in the operational model of RFOs.

The outline of the survey questions and categories was developed in response to the initial assumption of a lack of common definitions regarding participation actors and participation modes whenever "participatory practices" are mentioned. Within the core ethical practices of RFOs, the evaluation of funding schemes was detailed with these categories of participants. The same analysis grid was kept with all other categories of activities beyond evaluation, mainly knowledge sharing, consultation and program design. Also, beyond categories of participants combined with participation activities, other parts of the survey related to the ethical grounds sustaining RFO practices, ethics issues and legal compliance.

Since the limited data points would not have allowed for a relevant geographical distribution, the methodology was oriented toward a qualitative representation of results. The data synthesis serves to elucidate the parameters and factors of variabilities and to highlight conclusions regarding the current and future state of RFOs' ethical practices. Similarly, on the side of the expert interviews, results were pooled and elements on national contexts were taken into account whenever mentioned, but not included in the conclusions, in order to ensure anonymity and to maintain the same level of analysis for all inputs received.

Contextual specificities faced by each RFO were considered but not directly included in the survey. Keeping a stabilized list of categories[6] throughout the survey helped to better draw results from this analysis, while only stable categories of activities and participants[7] were considered, as common denominators across all cases. As a result, this survey is useful in testing the validity and coherence of these categories and criteria for any future ethics framework.

The general outline of the RFO survey involves a combination of categories of actors (participants) and of categories of RFO activities. The results are mostly of a qualitative nature, although in order to showcase the main results for each type of view, diagrams of a quantitative nature (aggregation of results) have been used. The choice of aggregation followed various types of combinations of actors and RFO activities, depending on the detection of major discrepancies: similarities were exposed and variations analyzed. All diagrams presented here were designed to showcase the survey contents in a comprehensive way, throughout the different parts of the survey outlined. The first and main part focuses on participants across the different types of RFO activities; the second part identifies beneficiaries of funding schemes; and the third part relates to ethics in funding schemes (the timeline of participation in R&I processes and the ethical grounds that are covered by RFOs). Although the first part is the main one, the other parts complement the analysis.

The diagrams showcase the gathered survey data in percentages, using a scale from 0 to 100 in steps of 10. The scale is color-coded according to the responses to each item in a ranking from 0 to 3, indicating the relative importance of participatory practices, with 0 when there is no such practice being employed and 3 when participation is the main mode of engagement.

Both current practices of RFOs and future expectations were included in the survey sent to RFOs and this divide was considered for the data synthesis, providing useful indications about expected variations in the future development of RFOs. Data synthesis was based both on data retrieved from the RFO survey and on data retrieved from the individual interviews – and the resulting contextual analysis for each RFO – in order to

6 See Appendix 1.

7 These stabilized taxonomies used for the RFO survey are indicated in the next section.

consider the variations in understandings and specificities that could compromise the stabilized analytical grid. The main results of the survey were kept as the backbone of the analysis, while qualitative inputs allowed for a more detailed analysis of ethical grounds and contextual specificities that shape participatory practices.

3.1.2. *Citizens' low involvement in R&I processes*

The sample of surveyed RFOs indicated the main categories of participants involved in R&I processes monitored by their structure as part of institutionalized participatory practices. To avoid broad unspecified answers such as "public participation", RFOs were asked to precisely identify through their activities which types of participants were involved across a diversity of activities, also specified. The findings indicated both the current state and the expected evolution, showing how RFOs identify priorities and needs for their future practices as related to participatory practices. The level to which ethical grounds are involved was also part of this survey, as analyzed through precise questions, and presented in this chapter.

3.1.2.1. *Main actors involved: experts' participation*

Overall, the results reveal expert-led governance of R&I in current practices of RFOs, which is also maintained in future expectations. The main actors involved in participatory processes across all types of activities are experts, governmental institutions and scientists.

In current practices of RFOs, these three categories of actors are mainly involved in activities of evaluation, knowledge sharing, consultation and program design. For future expectations, RFOs indicated the intent to maintain the predominance of experts, governmental institutions and scientists across all activities, mainly in program design and consultation, and to a lesser degree in evaluation, knowledge sharing and communication of funding activities.

The same main actors in participatory practices can be seen in future expectations of RFOs, which confirm the long-standing expert-led governance of R&I, as can be seen in Figure 3.2.

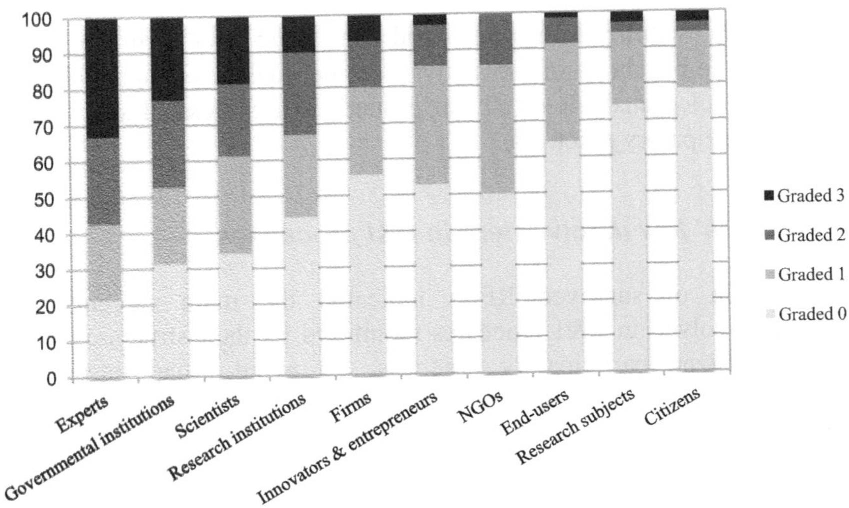

Figure 3.1. *Main actors in participation across all types of RFO activities in current practices (cumulative view/aggregation)*[8]

The prevalence of experts, governmental institutions and scientists was underscored by the aggregation of results across all dimensions of RFOs' activities. The following breakdown of cumulative results through detailed results for each of the three prevailing categories indicates the types of activities they are mostly engaged in.

As the first-ranked category of participants, experts substantially contributed across all dimensions of RFOs' activities: more significantly in evaluation, which is the main type of activity; secondly, in consultation and program design; and, third, in the technical monitoring of funding.

8 The 10 categories of participants are here combined with the seven dimensions of participation modes that were selected. The gradation (from 0 to 3) indicates the results obtained for "no" (ranked "0"), "yes" (ranked "1"), "yes to a significant degree" (ranked "2") and "mainly" (ranked "3"). The different categories of participants and participatory activities (and their definition for the purpose of this survey) can be found in the Appendices of this book.

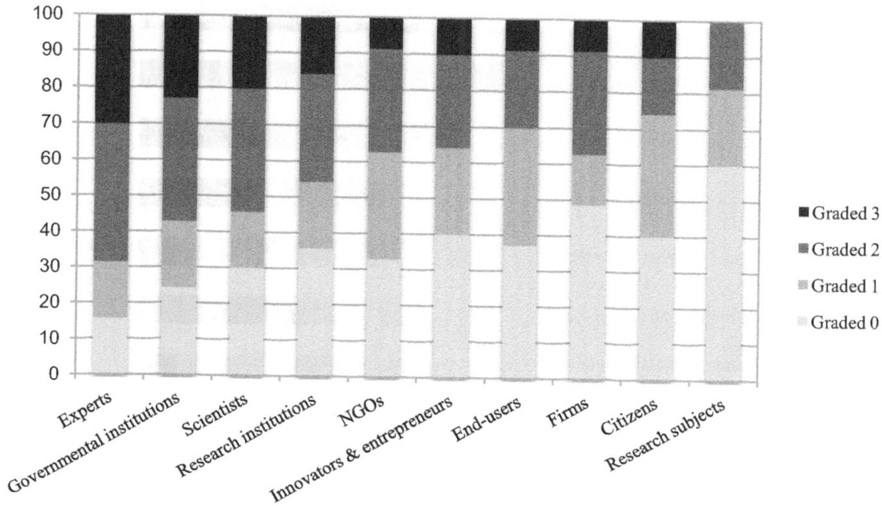

Figure 3.2. *Main actors in participation across all types of RFOs' activities in future practices (cumulative view/aggregation)*

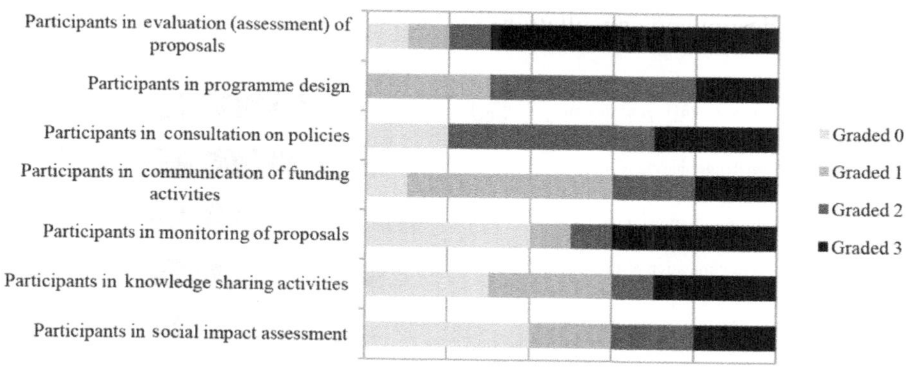

Figure 3.3. *Experts' involvement across all types of RFO activities in current practices*

Experts' engagement in the future activities of RFOs is expected to evolve stably as they are the main type of participants. Project evaluation is again the main area of involvement, followed by program design and, third, consultation.

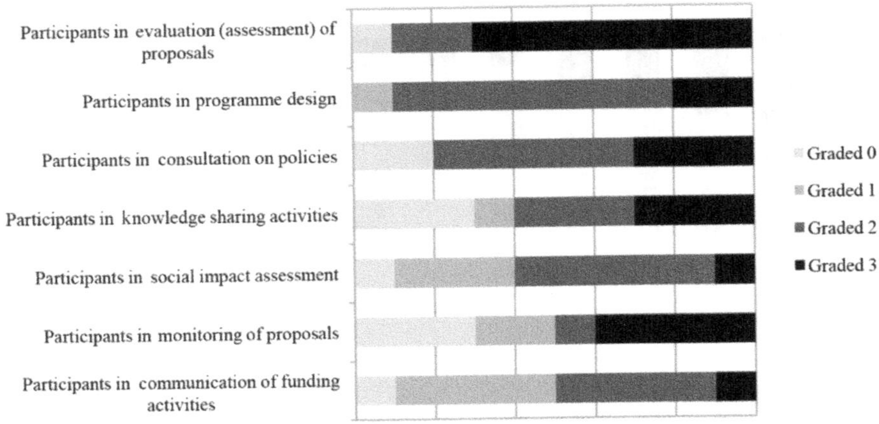

Figure 3.4. *Experts' involvement across all types of RFO activities in future expectations*

3.1.2.2. *Involvement of governmental institutions*

The second type of main actor involved across RFOs' activities is governmental institutions, which are mostly involved in program design, knowledge design and consultation.

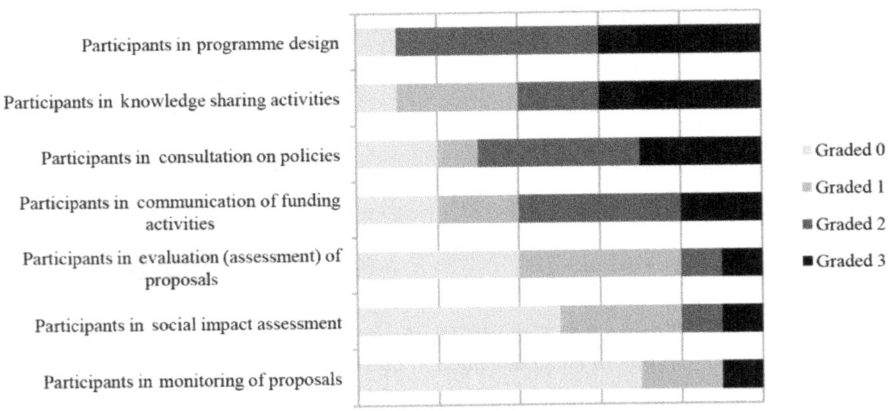

Figure 3.5. *Governmental institutions' involvement across all types of RFO activities in current practices*

Indications for expected future involvement show a result similar to current practices, with program design as the main activity, followed by communication of funding schemes and, third, consultation activities. Governmental institutions are a category of participants expected to evolve steadily in future RFO activities.

3.1.2.3. *The involvement of scientists*

As the third main category of actors in RFO activities, scientists are mostly involved in knowledge sharing activities, a category that is also maintained in future expectations. The other types of activities to which they contribute in current RFO practices are mainly evaluation and consultation, and in third position is communication of funding activities and program design, a distribution that is also expected to remain quite similar in future practices.

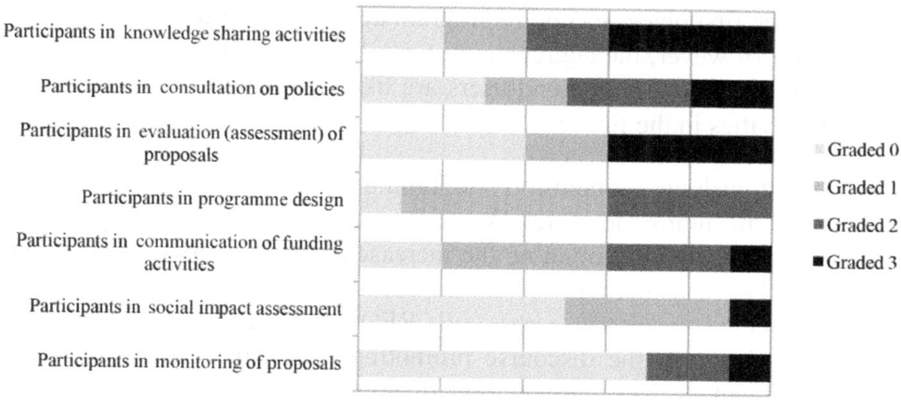

Figure 3.6. *Scientists' involvement across all types of RFO activities in current practices*

3.1.2.4. *The low representation of citizen participation*

As the cornerstone of participatory practices in publicly funded R&I at the EU scale, citizen participation is sustained by the democratic values associated with it, although the fine line between citizens and other categories renders the identification of public participation difficult. The citizens/stakeholders divide is not always clarified, while end-users have also to be distinguished as a specific sub-category of the two aforementioned types of participants. In the current taxonomy of participants, citizens and end-users were *a priori* expected to be at the core of participatory endeavors, both in current practices and in expected future practices. Yet, the results of the survey have shown that the role of these two important categories of actors is surprisingly peripheral, at the bottom of the overall distribution of participation actors. This result is also due to the broad definition of participation among RFOs, and to the fact that consequently all categories and all external participants were equally considered, without restricting some categories to certain types of actors.

The aggregation of results of the survey indicates that both citizens and end-users stand among the less involved categories of actors. Their current involvement in RFO activities occurs mainly in program design, knowledge sharing and evaluation activities, whereas future expectations indicated by RFOs locate the involvement of citizens and end-users mainly in knowledge sharing in the first place, second in program design and third in social impact assessment. However, the aggregation of results across all types of activities has shown[9] that citizens and end-users are likely to remain poorly involved in RFO activities in the future.

This astonishing conclusion stemming from the aggregation of participants in future activities is a contrasting feature in view of the discourses and policies promoting the increase of citizens' involvement and, more widely, considering social impact discourses. Citizens are an underrepresented category across all RFO activities: this indicates a discrepancy between the discourse promoting their participation and their actual involvement. The participation of non-traditional stakeholders does not seem to be particularly pervasive, and in the case of the present sample of RFOs, only half of them currently have programs involving non-traditional stakeholders.

The distribution of citizens across RFO activities highlights that their involvement is mainly in program design and in the evaluation of proposals:

Future involvement of citizens is expected to occur mostly in knowledge sharing activities; second, in social impact assessment and in consultation on policies; and, third, in program design activities. Although citizens' future involvement is poorly represented overall in terms of proportion across

9 See Figure 3.2 above.

expected levels of involvement, it has to be noted that RFOs mentioned the intent to re-adapt citizens' involvement in terms of types of activity, through a shift toward social impact assessment and consultation. This expectation seems to respond to the prevailing discourse on the need to involve citizens in decision-making processes and in ensuring the adequacy of actions with regard to their impact on society. However, in light of citizens' limited contribution, as less involved actors, there is doubt regarding the adequacy of means to achieve the indicated goal if their contribution is quantitatively limited across participatory processes.

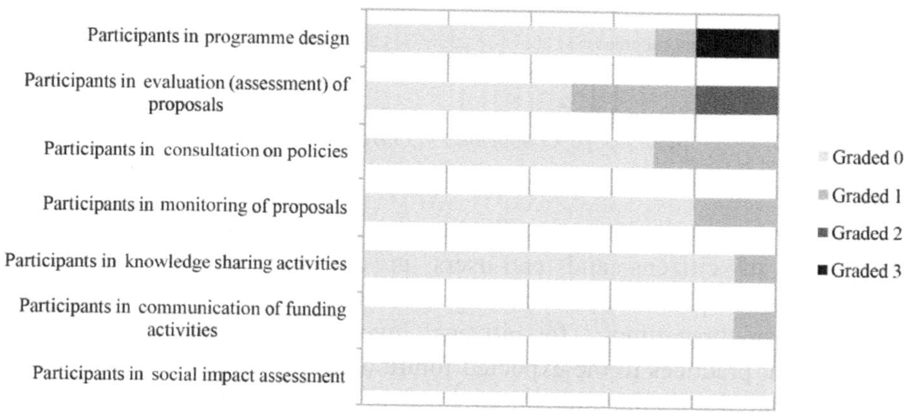

Figure 3.7. *Citizens' involvement across all types of RFO activities in current practices*

3.1.2.5. *The low involvement of end-users*

The same questions arise with the case of end-users, also ranked among less involved actors in the cumulative aggregation of results[10], both in current and future practices. Their current involvement focuses mainly on program design, while social impact assessment is poorly represented.

The prediction of RFOs regarding the future involvement of end-users indicates an expected prevailing contribution to knowledge sharing, program design and social impact assessment. Although these assumptions are significant in light of the expected growth, however, similarly to citizens, the low ranking in cumulative results tempers this progress.

10 See Figures 3.1 and 3.2 above.

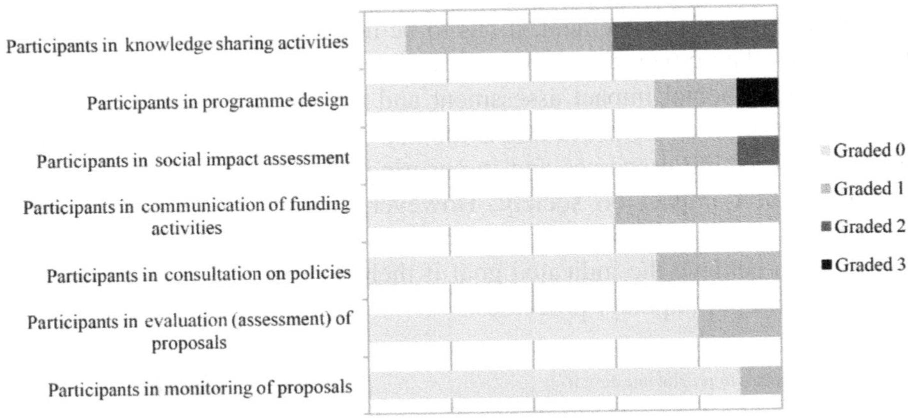

Figure 3.8. *End-users' involvement across all types of RFO activities in current practices*

Another approach that may alleviate these conclusions is the different proportion of citizens and end-users in current practices and future expectations, despite their low ranking. Cumulative results show an increase of more than three times[11] for citizens' involvement across all categories from current practices to the expected future ones. For end-users, the growth is more than the double the initial level of involvement[12], and this also confirms a substantial change. The remaining question deriving from this analysis is why citizens and end-users, who are both little involved in current participatory practices, are not expected to reach the first ranks of participants in future RFO activities.

3.1.3. *Prevailing types of participation*

The various dimensions of participation considered in this study refer to different ways of considering participatory practices, in terms of approach, purpose and outcome. These dimensions showcase an internal diversity that

---

11 From 21 points to 67 in the aggregation of results for citizens' participation, taking into account the degree of involvement (0 to 3), and the cumulative results across all types of RFO activities considered.

12 From 32 points to 71 in the aggregation of results for citizens' participation, taking into account the degree of involvement (0 to 3), and the cumulative results across all types of RFO activities considered.

can be analyzed either by type of participant across all sectors of activity or in a comparative approach of types of participants through each specific sector of activity. Apart from the dimension of social impact assessment, which has been previously considered, all other dimensions of activities follow a pattern of prevailing expert-led participation, both in current practices and in future expectations. As an illustration, the dimension of project evaluation reflects this pattern.

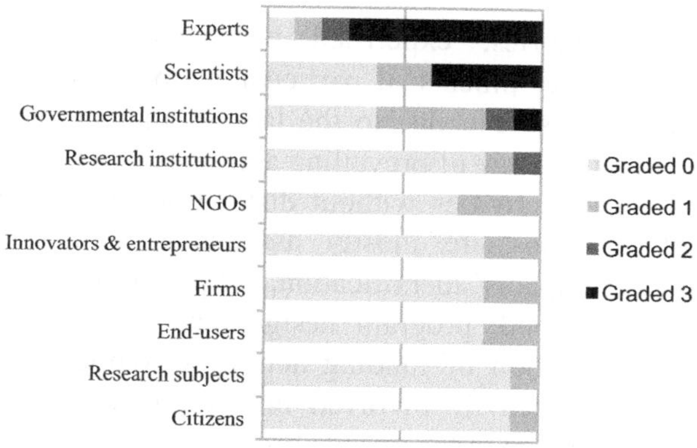

Figure 3.9. *Distribution of categories of participants in current practices of RFOs in evaluation practices*

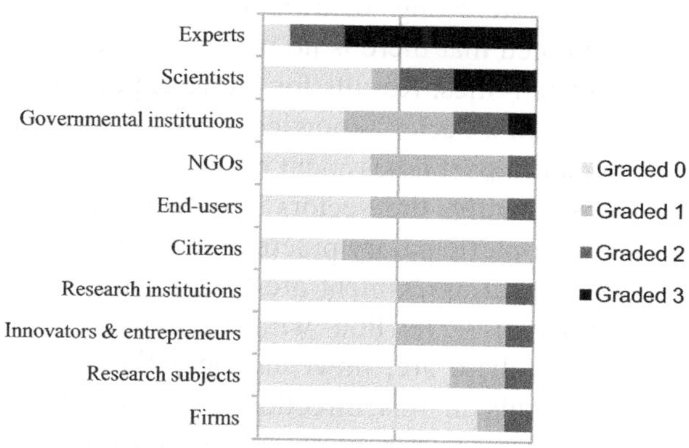

Figure 3.10. *Distribution of categories of participants in expected future practices of RFOs in evaluation practices*

In project evaluation activities, the main participants are usually experts, either internal to the RFO or external, depending on each RFO's structural means and its capacity to rely on external support. Scientists and governmental institutions are joining experts as a prevailing group of actors: this indicates that evaluation mostly relies on expert-led assessments. Even though underrepresented, citizens and end-users are growing categories in expected future practices, which is arguably a sign of the perceived importance of their inclusion in future developments of RFOs.

Across most categories, expert-led participation suggests limited flexibility in future developments of participatory practices in fields other than social impact assessment, due to the lack of motivation on behalf of RFOs. From the point of view of prevailing activities in RFOs' participatory practices, a ranking of activities without distinction of participants shows which types of activities are mostly associated with such practices. Regardless of the participants' identification, the prevailing sectors are: first, knowledge sharing; second, program design; and, third, consultation on policies and communication of funding activities. While these strands of activities are the most heavily involved in participatory practices, social impact assessment and monitoring of proposals are usually internally monitored by RFOs, without significant endeavors toward participatory processes. The diagram in Figure 3.11 showcases this ranking by sector of activity.

In addition to this interesting distribution of prevailing sectors of activities, it should be noted that there is no particular evolution expected on this side in future RFO activities. Results for future expectations come to the same conclusion for the first sector (knowledge sharing), with some minor fluctuations, while social impact assessment remains lower in comparison to other categories. Consequently, the sectors in which RFOs are hoping to successfully develop their participatory practices seem already settled. While evaluation and social impact assessment are traditionally referred to as the main foci of participation – in line with the legacy of participatory technology assessment – there are, however, only limited expectations of reaching participatory goals in this direction. What do these conclusions signify, as they actually are counter-intuitive in light of RRI guidelines and the scope of PRO-Ethics? Additional elements stemming from the legal, structural and ethical background of national RFOs and R&I governance, as well as experts' insights, may bring more elements to understand these first assumptions and conclusions.

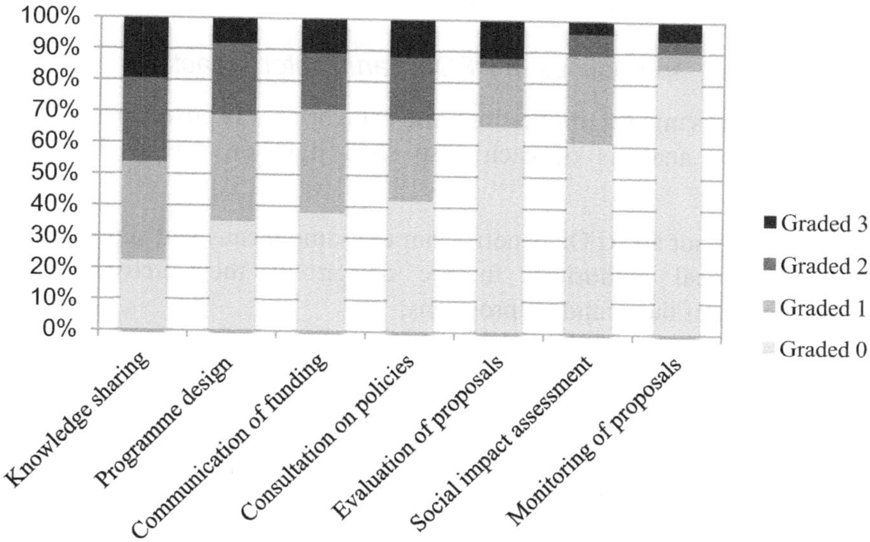

Figure 3.11. *Ranking of current RFO activities according to the degree of participatory practice (regardless of distinctions between participants)*

Regarding the timeline of participatory practices covered by RFO activities across project timescales, most RFOs have indicated that in their current activities, they involve participants mainly before R&I, and also, to a lesser degree, during R&I processes. The *ex post* (after R&I) involvement of participants is less developed in current RFO practices. This tripartite determination of R&I processes across time sketches some orientations but cannot possibly reflect a precise correspondence to specific strands of participatory practices, due to the complex entanglements of participation activities and project timelines. This aggregation showcases the overall focus of RFOs, regardless of the specificities of participatory processes. Expected developments in future practices indicate an increase of pre- R&I activities to the maximum level and an increase of post-R&I activities reaching the same level as during implementation. The intentions of the surveyed RFOs are mostly to better anticipate and to be better equipped in impact assessments.

3.2. Leveraging the funding of R&I for a novel approach to participation

3.2.1. *Context-sensitive specificities*

3.2.1.1. *Varying approaches in RFOs' participatory practices*

Although the same set of questions and categories was posed to all RFOs, the background analysis of each case shed light on several levels of discrepancies:

– *structural*: not all RFOs benefit from the same means and support from governmental/local authorities in the design of their activities and implementation of their funding programs;

– *legal*: national laws are introducing specificities in addition to compliance with European laws;

– *ethical*: not all RFOs allocate the same importance to ethics and ethics procedures.

These general assumptions stem from the qualitative data gathered from the RFO survey, both from the survey questionnaire and the follow-up interviews that served to go further into details on the respective backgrounds RFOs operate in. The exposed discrepancies have to be taken into account in future guidelines and frameworks for public funding schemes in R&I. RFOs' current practices and expectations for future practices are intrinsically linked to contextual specificities whose diversity suggests the European landscape of RFOs is far from homogeneous and that any attempt to improve the gaps will be subject to debates on the way to filling in the blanks.

The impact of participatory processes is particularly context-sensitive (sector and context). This correlation between particular sectors or contexts and the impact of participatory processes might be salient for instance in R&I sectors for which participation has a significant leverage, such as: gene technologies, energy, environment (climate) and the health sector. In these fields, participatory processes have had great influence so far, and notably in decision-making and the exploitation of new technologies. Although there is a need for more consensus in society and more societal dialog around R&I and technologies, this is not always implemented – some sectors, such as AI, are particularly in need of societal dialog. Also, healthcare could be mentioned as a sector where the involvement of participants (e.g. patients) can directly increase the relevance and the utility of the outputs of R&I and ensure the inclusion of individuality and differentiation whenever there is technology with end-users. Negative impacts could arguably occur in the event of an inclusion of industry interests in the development of national R&I strategy, or when participation slows down R&I processes. The impact of participatory processes is also dependent on the connection to decision-making and the specific parameters related to the methods and structure of the initiative. As the impact can be messy and create concern and uncertainty, participation has to be conducted in a clear way in terms of structure and responsibility. Non-effectiveness or even counterproductive results can occur if the organization of the participatory process is not adequate, and the conditions not properly developed. In addition, some common issues on behalf of participants, such as lack of confidence and fear of evaluation or even competition, may hinder dialog and idea sharing during the participatory process. Also, as the final decisions deriving from participatory processes do not always reflect the views that have been expressed, there is a need for strict accountability processes and transparency, for instance by including participants in management and decision-making. As a counter-argument, it might be objected that the impact of participation does not depend that much on the importance of the context (or sector), but on the quality of expertise and understanding of expectations in each process. Understanding the social context is a key issue in participatory R&I, because it has an impact on stereotypes, biases and general beliefs that are not necessarily evidence-based, although deeply rooted in societies. From this perspective, having broad participatory processes would imply an attempt to capture and understand various voices as different realities.

Evaluation procedures are another pillar in this study that proved quite divisive both in terms of RFOs' current practices and at the level of expectations and views on the future development. The ethics review – as a formalized standard procedure – is not common to all European countries. Some RFOs do not endorse standard ethics procedures, unless they are a compulsory component as part of a European funding scheme. While this indicates the need for enhanced dialog and cooperation at the European level in order to tackle the diversity of procedures, it could also be argued that ethics reviews should not be applied to all R&I sectors, and there should be distinctive criteria to successfully support this divide. From this perspective, some open questions arise as to how a high standard could be achieved, and how to counteract the sensitive dimension of guidelines in some contexts, including rejection, which should be faced in order to be softened.

The idea of reaching unified guidelines and implementing them could prove difficult, given differences in approaches. If the same standard of ethics review is adopted across countries and sectors, it could however be timesaving for basic ethical dilemmas. This minimal ethics procedure cannot however overshadow ethical deliberation, the monitoring of projects and the decision over a project's eligibility to receive funding. Ethical evaluation can be useful in helping a project team if the project is scientifically sound and deserves to be funded, but denying funding because of ethical reasons, without further examination, could be counterproductive. Also, enhanced ethics procedures are needed in some specific sectors, such as the medical sector, where ethical deliberation can intervene on top of ethics reviews, but where standard ethics evaluation procedures ensure a standard of care and shared responsibility. Ethics reviews could be perceived as a baseline, ensuring the basics or non-negotiable parts reflecting principles are covered, prior to considering deeper insights. Ethics reviews could be considered as the main tool ensuring research integrity as a component of R&I quality, preventing harm and malpractice related to research or data management, for instance; as such, guidelines going in this direction could be uniformly compulsory. National adaptations following general principles could intervene to ensure evaluations are context-relevant.

3.2.2. *Expectations and legitimacy in the funding of R&I*

3.2.2.1. *Expectations regarding participation*

Beyond what the scientific and gray literature mentions on the topic, the expected outcomes of participatory practices are not only relevant from the side of theory. The way institutional actors organizing participatory initiatives relate to the outcomes is all the more important, as it is intertwined with the design method and the resources they are willing to allocate to either implement it in the best possible way or to improve these participatory mechanisms. As part of the RFO survey, questions pertaining to the expected outcomes and the perceived legitimacy of participation were asked, in order to indirectly gather information on the institutional perception of the adequacy of participation and the underlying ethical values associated.

Overall, data from the survey have shown prevailing concerns regarding legal compliance, as well as expert-led governance of R&I across main RFO activities. The limited difference between current practices and future expectations indicates a narrow capacity for change both in the contribution of ethics and in the way participatory processes are designed. The interviews with RFOs have allowed for a close-up view on their aspirations and the constraints they are facing in their future development.

RFOs were asked to answer the question, "what do you expect from participation in your current activities?" and indicated various dimensions as part of their current endeavors toward participatory practices. Their expectations can be summarized around four main categories:

– *social impact* (the improvement of both the social acceptance and social impact of R&I, as part of RFOs' current participatory practices)[13];

– *operational effectiveness* (anchoring participatory practices in a result-oriented approach to funding schemes by delivering better-informed program design, better funding allocation – and identification of funding targets – better identification of topics and better expertise in projects' evaluation);

– market-oriented outcomes;

– *public awareness-raising* (ethics readiness of the public).

Regarding their expected future activities, RFOs suggest a turn in favor of social impact throughout funding processes and outcomes. RFOs' future expectations express the prevailing importance of social impact and of operational effectiveness as part of social impact implementation. Most of the RFOs indicated that they expect improvements in increasing the societal impact of their activities as a timely response to evolving societal needs. The improvement of social impact is considered by RFOs at the levels of:

– better inclusion of societal needs and challenges in funding schemes;

– moving from participation for fair project evaluation toward co-creation;

– program design and topic identification to bring programs closer to society (providing solutions to target groups' needs);

13 This entails, for instance, working in the direction of end-users' involvement in order to reach better societal acceptance of new technologies.

– involvement of citizens in social impact assessment;

– support of state-of-the-art R&I sectors in a sustainable way.

Ultimately, such endeavors should contribute to bridging the gap between science and society, leading to more consistent approaches to stakeholders' involvement across all types of actions undertaken by RFOs.

3.2.2.2. *Considerations regarding the legitimacy of participation*

The expected outcomes are only one side of the coin, the other one being the perceived legitimacy of participatory practices and the views on the effectiveness, the adequacy and the quality of the outputs.

Another strand of the qualitative investigation in RFO activities related to the nature of the legitimacy of their participatory practices. The question asked was, "what is the legitimacy of participation in your current R&I activities?", as an open question allowing for a wide array of approaches to participation. The answers to this question indicated a variety of sources of legitimacy, which can generally be summarized as follows:

– project relevance;

– fairness in funding allocation;

– external legitimacy;

– insufficient tools to implement legitimacy.

While the first three answers indicate some criteria perceived as a fair justification of participation, the last one is a negative answer, as some RFOs struggle to find the sufficient resources to implement participatory initiatives as they would ideally tend to do, whether in terms of configuration, scope, participants, structural means for monitoring, proper connection to decision-making and influence or merely the capacity to undertake a follow-up covering all stages, from the inception to the *ex post* assessment.

Several RFOs have underlined the connection of participatory practices with project relevance, as participants are a source of information benefitting RFO activities in R&I funding. The legitimacy in this case stems from the assumption that project results will better contribute to tackling specific challenges through the selection of appropriate stakeholders. Also, initiatives such as partnering with expert organizations to deliver citizen participation activities (such as convening public dialogs) were mentioned as an important step in the involvement of relevant stakeholders in the R&I processes. The legitimacy built on the relevance of funding projects (better-designed and -oriented projects) also ensures that projects will result in more desirable outcomes.

Another pillar sustaining the legitimacy of participatory practices is the fair and efficient allocation of resources, which responds to the need for efficient allocation of the scarce public resources for R&I. The selection mode of participants is a key point, as it has been suggested that participants should be selected according to each project's specificities. When participatory practices are an integral part of funding programs, and procedures have been previously established, such legitimacy in the fairness of funding allocation can be attained.

Even though legitimacy is usually sought in the way participatory approaches are constructed and implemented, some RFOs indicated legitimacy gained from external recognition or support. In these cases, RFOs seek legitimacy in the institutional support received from governmental authorities. Although ethical grounds do not intervene in this form of legitimacy, this external support can serve for some RFOs as a sufficient indicator of the value of their ethical practices.

Not all RFOs have at their disposal the resources, the tools or the expertise to implement participatory practices, and these difficulties are another case of legitimacy that is being sought but not achieved yet. The lack of sufficient legitimacy can be due to doubts or difficulties in the implementation of participatory practices: these were reported by RFOs as impeding factors (participation as opposed to research integrity; increased costs; doubts about the added value of participation). Another cause can be found in the lack of inclusion of participation in RFO strategies when there are no data reported on participation activities in the R&I funding activities conducted by RFOs, either due to lack of interest or lack of the methodologies to develop such strategies.

Future expectations for fostering legitimacy in participatory practices reflect the same ideas, with some additional features on the side of the development of local innovation ecosystems and the development of policies ensuring partnerships and programs to actively implement equality, diversity and inclusion goals. In addition, as a legitimacy criterion, it was mentioned that public policies should be evaluated from the point of view of their added value to society (citizens' interests and objectives). Also, it was noted that legitimacy can be found in the support of collaborations aiming at reaching a critical mass in view of advancing the global value chain. All changes seeking better legitimacy should be reflected in changes in regulations on science, higher education, technology and innovation.

3.2.2.3. *Potential conflicts and tensions*

Legitimacy conflicts between ethics and participation will keep arising as long as the bridge between the two dimensions is not framed by consistent guidelines and quality criteria. Diverging practices and approaches to participation – involving a variety of participants through different selection processes and for diverse types of activities and outcomes – share the same difficulties in settling appropriate practices that are ethically sound. The resulting tensions between the ethics and participatory processes of RFOs generally indicate the following dimensions:

– conflicts of interest and biases;

– accountability toward participants;

– economic or scientific development in conflict with ethics;

– diversity of ethical priorities[14] across R&I sectors and contexts.

Furthermore, several RFOs have indicated difficulties in tackling conflicts of interests in the involvement of external stakeholders: as they each represent their own interest[15], considering their views can result in biased final decisions. In other words, biases may arise when participatory processes are dominated by specific groups[16]. This requires handling the

14 RFOs may have different levels of priorities for ethics depending on programs' specificities: for instance, programs on emerging technologies can be anchored in public values, whereas funding programs supporting early-stage innovations might not even have an explicit consideration of ethics. Even within a single RFO, different projects and teams lead different objectives and sometimes with no consistent approach to ethics and participation, as was reported by RFOs.

15 Multiple interests can be at stake for a single actor (a citizen, for instance, can have several allegiances).

16 Deliberative models of participation aim to tackle this issue.

tensions that occur when deciding how to appropriately identify stakeholders and which category to place them in the course of the participatory process.

Developing public engagement may also be in conflict with ethics, since accountability has to be further developed in R&I governance. Regarding citizens' involvement, the danger of exploiting the willingness of citizens to collaborate looms on the horizon as long as participatory processes are not framed by ethical guidelines. In addition to this, a tension between ethics and participation lies in the uncertainty about how well-informed participating citizens are[17], depending on the selection mode and the way they are involved in R&I processes. Citizens' involvement has to comply with ethical standards (i.e. informed consent) and ethical issues have to be considered throughout the whole lifetime of the project, including in exit strategies at the end of the project. Tensions in the involvement of vulnerable groups may also arise when participating citizens have needs and expectations that are disregarded during their involvement. The need for accountability toward participants – either citizens or stakeholders – entails that RFOs should consider all voices and not choose which inputs will drive their activities and exclude others.

Tensions can also arise from diverging rationales between scientific or economic development and ethics, especially in the event of the imposition of a stringent framework for ethical R&I. Considering beneficiaries of funding schemes as a type of participants, tensions in their participation can arise when their scientific readiness is not followed by an ethics readiness. This latter can impede the implementation of the project, since in the course of the project's development an insufficient ethical approach may suspend the project and its participation in the funding scheme.

In priority setting, RFOs have varying approaches to ethics, not only depending on specificities of certain sectors and contexts (national or local priorities), but also internally, as different programs of an RFO may place different levels of priority on ethics: for example, programs on the governance of AI may focus on the connection to public values, whereas other funding programs supporting early-stage AI innovation may have no explicit consideration of ethics. The discrepancies in priorities reflect the variety of objectives, as there is no consistent approach to considering ethics

17 From an ethics perspective, this is an issue worth considering for all categories of participants (beyond citizens).

and participation. Another issue on the priority-setting level has been noted in the difficult mobilization of the business sector, which does not always allow for a broad definition of ethics and participatory processes.

Future expectations of RFOs regarding potential tensions between ethics and participation are globally the same as those perceived in their current practices, and it was mentioned that tensions can be alleviated through awareness of ethical issues, which increases along with the improvement in participation efforts and when guidelines and regulations gain greater implementation. There is also a need to enhance the capabilities of RFOs to appropriately identify stakeholders and target groups and to adapt participation accordingly: from this perspective, stress has been put on participation in evaluation processes, where consistency is not easily achievable at the EU level. Also, the uncertainty of the opposition between the commercial dimension of projects and their social impact was mentioned as an important concern. At the same time, accountability and responsibility toward participants are thought to remain important issues in future developments of RFOs across their participatory journeys.

The place attributed to ethics is part of the contextual specificities of RFOs, as most of them rely on different ethical grounds and hence different understandings of – and references to – ethics in their activities. Part of the RFO survey was dedicated to ethical grounds in order to determine on what grounds the ethical dimension of participatory practices is built on in each case. Ethical grounds of participation indicate the nature of the involvement of actors in RFO activities, justified and legitimized through diverse perspectives. Several approaches to ethics can be identified, among which three main categories are as follows:

– ethical concerns (e.g. research integrity);

– ethics broadly understood, according to moral values or to ethical theories (principles, consequences, virtues such as care);

– compliance with ethical review processes, RRI or national or European laws.

Currently, legal compliance takes over ethical concerns regarding the ethical grounds of RFOs, both in terms of the definitional and procedural aspects of ethics. The main ethical grounds indicated by RFOs are: first, compliance with national laws; second, compliance with EU laws; and, third, compliance with moral values.

The main issue with this is that the distinction between ethics and legal compliance might be perceived differently between RFOs. Moreover, the practical implementation of ethical and legal frameworks within RFO practices (e.g. ethics review processes) is in no way homogeneous and is subject to different interpretations and adaptations in national contexts.

3.2.3. *The impact of participatory practices*

The main added value and main justification of participatory processes in RFO activities can be outlined in three equivalent categories:

– a focus on quality and value of procedures;

– a focus on connection with societal needs;

– functionality.

An important focal point of the main added value and the main justification of participatory processes relies on the values and the quality of procedures, as for instance with the gain of broader and deeper understanding and more legitimate and fair processes. These values reflect the idea that funded R&I should benefit everyone: technology concerns everyone in society and, therefore, broad engagement should occur in the processes, development and implementation of science and technology. It could be considered that documenting the ongoing reflection during participatory practices is a key element for the added value in that a well-documented participatory mechanism can be more than a temporary feature. Another value could be perceived in the sustainability of decisions, which is usually better in cases with greater participation, where diverse arguments have been heard and taken into account.

The source of the added value and justification of participatory processes could also be located in a better science/society connection, hence better R&I adequacy with societal needs. Participatory processes are contributing to the challenge of defining what people want, demonstrating their needs. By bringing pieces of reality into R&I processes, participants help better grasp processes and align these latter with a more relevant and legitimate society. Addressing grand societal challenges could also be another main added value, grounded in the idea that including people from policy-making spheres, or citizens, would ensure a better response to societal challenges and better decision-making on such topics. Also, participatory processes defy the common approach based on regular statistics in R&I, where there is not enough scientific evidence-based information and not much diversity in terms of data used for decision-making. Participation improves this by combining as much diverse data as possible, and with the contribution of diversified viewpoints in reflection, as an ongoing process: this tends to be superficially dealt with in current RFO activities, according to some experts. Managing the R&I policy with these participative processes results in the main added value of connecting RFOs closer to the reality of societal needs, ultimately achieving a better connection with the reality of both our societies and lives.

Other key factors for the effectiveness – the added value and the justification – of participatory processes could be considered on the side of the practicalities of R&I processes, specifically in the better functionality of R&I processes. Thanks to participation, innovation in general gains better acceptance and integration in everyday life, in a more functional process. Also, with the integration of perspectives that would have been missed without participation, having R&I better fitted with contexts and challenges ensures better value for public investment in R&I. With the need for more market-ready products, as well as better functionality and acceptance, participatory practices can be oriented in the direction of the avoidance of collateral damage (technical or political) from innovation. Participation supports R&I growth inasmuch as there is a benefit for the scientific field to draw from social and human sciences in order to organize information retrieved from society, so as to better define and demonstrate the needs, and to make sure R&I does not address small groups only.

Among the multiple objectives motivating participatory processes in R&I, some main goals and outcomes should be prioritized: value-oriented goals or efficiency indicators. Value-oriented goals and outcomes, such as democratization, consider that a push toward participation could counterbalance expertise and technocracy, provided the actual input from citizens is taken into consideration. Also, openness could be an expectation from participatory processes, through the collection of diversified data and expertise, along with a culture of good debate. Another goal could be the idea that innovation should be acceptable for everyone and not only for those who create or organize innovation: therefore, the wider the participation is, the better the acceptance and legitimacy can be. Another target of participation could be to increase transparency, since although information is easily accessed in our societies, transparency has still not improved enough. Furthermore, another motivation could be found in the involvement of citizens participants in all R&I stages, without narrowing down the involvement to a partial contribution but aiming at full participation, bringing insights from society at all R&I phases and following different modes of participation.

Considering efficiency indicators, from this perspective, participatory practices should aim at increasing the quality, relevance, acceptability and applicability of R&I outcomes, such that they can have a real impact on innovation and policy development. The quality and efficiency of the final outcomes of participatory R&I can be identified in better design of science and technology. Other views on the matter focused on the economic efficiency of R&I outcomes: the goal of a strong economy can be better achieved through participatory processes. Also, another suggestion is that in an open R&I configuration, public involvement can help shape a strong ecosystem, ensuring equality and diversity. In addition, market and policy values can guide R&I participation toward market-ready and efficient outcomes.

Some indirect or unspecified outcomes of participation are also part of this global picture, as some objectives are not always directly related to R&I product or service outcomes. A main goal could be to consider the benefits and acknowledgment of the participants' contributions, addressing privilege and balances in global R&I architecture through more inclusive processes and decisions. Such kinds of benefits could range from monetary benefits to prestige sharing, whereas acknowledgment could also take various forms – as long as the main goals and outcomes are built around participants' contributions. Another view also diverging from participation goals anchored in R&I products is the idea that participatory processes have unspecified outcomes and goals, following the undetermined nature of R&I. This view implies that being open and making interpretations of results that could be different from expectations is part of a sound scientific process, which means that the outcomes of participatory processes cannot be determined beforehand.

3.3. Addressing the needs of research funding organizations

3.3.1. *Standardized approaches and the European discrepancies*

How can ethical participation be achieved? Reaching this goal would require taking stock of RFOs' needs and supporting the improvement of existing ethical schemes whenever they are effective. In concrete terms, this implies considering: the needs of RFOs and the way ethical appraisal schemes (encompassing both ethics reviews and regulation compliance) should be enhanced with participation; the current and future needs of a participatory dimension among project partners' activity; and the incentives and the means that would allow for the proper development of participatory schemes. Considering existing tools – such as ethical and legal frameworks within RRI and ethical review processes – the question of what would be a useful direction for new interaction modes/novel participation configurations remains to be specified according to the context (i.e. sector, scope and impacts, potentially affected publics and ability to address ethical issues and participatory options). In addition, cultural specificities need to be considered since not all public funding schemes are equipped similarly in terms of available infrastructure, financial means and the capacity to develop ethical expertise. As a result, not all RFOs address participatory processes in their current activities to the same extent or with the same approach. The very idea of "participation" is a blurred landscape of unclear terminologies, views, goals and approaches. The foundation of this study arises from this difficulty, as it is aimed at confronting this messy field by proposing some tentative categories and distinctions that could serve to better frame the different layers of approaches and understandings of participation.

The case study of RFOs in Europe explores the different characteristics of standardized participatory approaches. Considering eight RFOs and two other entities (an international network and an expert body that also conduct substantial activities as RFOs), they all reveal different operational models, different legal frameworks and also different ways to relate to ethics and participatory practices accompanying the geographical diversity of this sample.

The first major discrepancy at the EU level relates to the way each national RFO relates to ethics, from the point of view of procedures and expertise in this field. Within the considered sample of ten surveyed RFOs, only one declared they had a formalized ethics procedure (ethics reviews) as part of the evaluation process of funding schemes. In addition, a second RFO had a formalized ethics procedure, but only for funding programs supported by the European Commission. Also, this RFO that follows a formalized ethics procedure in its projects' evaluation processes is also the only one in this sample of RFOs to have an ethics committee. This may indicate the interdependence between a formalized ethics procedure and the development of specific internal expertise in ethics inside the operational model of a given RFO.

With regard to the way participatory procedures are structured, the focus on specific types of participants is similarly developed, mainly with external experts, institutional partners (public bodies, firms), citizens and the greater public, in unidirectional participatory forms (knowledge sharing, communication and information).

A case-by-case analysis of each RFO revealed the lack of tools and means to further develop participatory methods and effectively address ethical issues. The following general needs and expectations were reported:

– the need for indicators for participation;

– the need for indicators for social impact measurement;

– the need for means to develop the ethical dimension through a focus on the ability to exchange views;

– the need for better ethical follow-up through the creation of a pool of experts;

– the need to elaborate the ethical dimension and participation further through program/call design and also to expand consultations with the civil sector;

– the need for guidance on ways to handle ethics issues, especially for key sectors that may present ethics issues, such as medical R&I, AI and autonomous vehicles;

– the need to develop consultation activities involving citizens and target groups;

– the need to broaden the scope of people and organizations involved in activities.

The future evolution of ethics in RFOs' operational models is torn between the need for better ethical monitoring and expertise and discrepancies in ways of defining ethics' embeddedness. The ethics review process is not standardized/formalized between the different European countries, except when RFOs' funding programs are directly supported by the European Commission. However, some RFOs have pointed out that ethics reviews might be superficial if only considered as a "checklist" and if they are not followed by specific assessments related to the nature of the call. Therefore, ethical dimensions could also be considered through dialog and targeted monitoring, further to formalized procedures that do not necessarily fully cover all ethical concerns or guarantee the ability to exchange views.

Whether structured or – as in most cases – indirectly dealt with, ethics is not equally handled and not all RFOs are equipped with the means and the expertise to ensure the ethical implementation of R&I processes beyond legal compliance. Also, the difficulty in implementing participatory processes is part of the insufficiently developed ethical dimension, as participatory modes, actors and outcomes are not substantially analyzed. Most RFOs seem to conduct participatory processes without labeling them as such and without having the tools to develop a strategic plan for such practices. The difficulty stems also from the various types of activities where participation can be implemented, either at the program, project or RFO (general) level, to summarize the various activities that have been considered in this survey.

In addition to all these gaps and expectations, some RFOs have indicated their commitment to enhance participation and inclusion in R&I processes through bottom-up principles, by setting standards that would allow for such autonomy. This balance is still to be defined, as the lack of indicators and criteria to develop purposely designed participatory strategies either in top-down or bottom-up approaches maintains a blurred field.

	Variabilities	Prevailing features
Operational model	Unequal relations to national governments and unequal support from governmental bodies	N/A
Approach to ethics (procedure and expertise)	Structured approach or lack of structured approach	Formalized ethics procedure and internal expertise in ethics (permanent ethics committee) are rare
Approach to participation	Structured approach or lack of structured approach	Main actors involved: experts and institutional partners Citizens' participation often limited to dissemination and knowledge sharing activities In addition, RFOs indicated participatory activities conducted without being labeled/identified as such

Table 3.1. *Variabilities in the operational models of European research funding organizations*

3.3.2. *Ethical grounds meeting public scrutiny*

Reaching fair, efficient and legitimate participatory frameworks in R&I requires defining the main quality criteria of participation, as indicators of good practices. The quality of participation is not similarly perceived: the focus could be put on values, on processes or on results. Overall, defining an *active* and *ethical* form of participatory process is the challenge at stake here.

For quality criteria related to values, these do not differ much from the dimensions previously examined such as legitimacy or goals. As a quality guarantee for ethical participation, notions of justice and equality are usually connected to this view, in terms of citizen empowerment and representation. This is based on the idea that R&I should be for the greater public, with an underlying normative view anchored in social justice. Although quality criteria could be considered from case to case, regular and deep engagement is necessary, along with control mechanisms making it possible to stop R&I processes when decision-making power is too distant from public engagement. Overall, several notions could be associated with these values, such as: accountability, transparency, openness, inclusiveness, honesty, feasibility, relevance, applicability and acceptability. A common idea behind these values is that the challenge is to create enabling environments allowing as many people as possible to express their opinion – considering inclusion as the main quality factor – thus counteracting power relations and structures. Having inclusive participatory R&I processes open to arguments enables trust building and knowledge building[18]. Although designing such horizontal relations takes time, some of these values – such as accountability – could be made mandatory: this would ensure that participants are not only transmitters, and that participation is promoted as a democratic renewal. Such values prevent a superficial mobilization of citizens and stakeholders and the risks it entails of loss of energy, time, hope, expertise and beliefs, thereby triggering serious negative impacts and democratic depletion. It has been raised that there might be a lack of maturity in R&I policies at this level in some national contexts whenever it is not acknowledged that the reply issuing from participation is as important as the production of R&I processes.

Quality criteria of participation could also be process-focused, anchoring quality in dialog, and in particular financial and structural support allowing this dialog, since ethical considerations and decisions need time and financial means, whereas dialog is often considered as an addendum to R&I processes. As part of this re-shaping of processes, genuine contribution from participation implies that suggestions brought to the table are integrated or taken into account, otherwise participatory processes remain at a passive stage, which leads to the devaluation of participatory processes. Also, additional elements that have been mentioned as part of process-focused quality criteria may need better relationships between different stakeholders and citizens involved in R&I processes, in order to seek for novelty without complicating societal issues.

Result-oriented quality criteria are another strand, pointing to a better identification of what a good outcome should be and how an R&I program could be of benefit for more people. This is also a matter of better framing of benefits, identifying their reach regarding policymakers, innovators, citizens and stakeholders participating to R&I. A related quality criterion would be the ability to measure the outcomes at each stage, which requires precise indicators. The usefulness of R&I results and the efficiency of processes can be considered as the main quality criteria satisfying both R&I organizers and participants: if all participants, members of RFOs, citizens and stakeholders

18 The design of enabling environments for inclusion is, however, an open question since several approaches (e.g. for participants' selection) compete, as has already been mentioned.

participating in R&I projects agree at the end of the project that it has changed their point of view and contributed to them learning something, the whole participatory process could then arguably be defined as qualitative and productive. Closer to social impact values, this result-oriented quality criterion implies that it is crucial that participants feel appreciated and that the participants' arguments make a difference in the R&I process.

Grouping together all input retrieved from the interviews with experts has made it possible to underline the main strands of the suggestions and viewpoints for each of the topics considered. This synthesis and analysis summarized the arguments that were gathered, taking into account the commonalities and the quantity of similar views.

3.4. Toward an ethics framework

3.4.1. *Finding common paths and compromises in R&I governance*

The case study of RFOs presented highlighted several main needs, which can be summarized as follows:

Key factors to be integrated in an ethical approach to (institutionalized) participation:

– common (and stabilized) categories of participatory activities;

– common (and stabilized) categories of participation actors;

– proper assessment of the available resources and capacities for the design and implementation of participatory activities;

– availability of ethics expertise (internally or externally to the given institution/organization);

– identification of the added value of participatory activities.

Box 3.1. *Critical features for an ethical framework in institutionally led participatory activities*

What could help improve the governance system? How can easy access to participation be achieved, with a representation of all societal groups relevant for each specific case? How can quality control mechanisms be implemented and the ethics of participation addressed further to regulatory and institutional frameworks? What is to be mainly expected from the participation of citizens and stakeholders? In the absence of a clear identification of the ethics of participation, how can practices and new initiatives be considered in a comprehensive framework?

In the absence of a common approach regarding the ethics of participation and given the diversity of parameters and contexts, there is no single definition that could encompass the dimensions outlined so far. The main challenge to be tackled by R&I institutions is probably the use of a stabilized taxonomy of participatory practices. This could provide a common reference and the possibility of organizing participatory practices following a thorough understanding of the configuration choices (participants, activities) and the underlying ethical dimensions.

Within the overall RRI objective, rather than a policy element, participation could be seen as a guidance tool to be re-defined through precise framing of innovation processes:

> Because innovation is an inherently complex and dynamic social process, there is tremendous value in connecting theory and practice. Indeed, effective policy requires policymakers to have a comprehensive understanding of what might work in theory and what is working in practice. (Bogers et al., 2018, p. 7)

The European Report on Research Ethics mentions the connection of ethics with participation, the latter being embedded by definition: “the way research ethics is interpreted at the European Commission, aims to be collaborative and constructive” (European Commission, 2013, p. 3). In the related gray literature, some blind spots raise the question of how to reach ethical participation. Answering this question requires taking stock of the needs of RFOs, as well as the way ethical appraisal schemes (encompassing both ethics review and regulations compliance) should be enhanced with participation; the current and future needs of the participatory dimension among project partners’ activity; and the incentives of and the means that would allow for a proper development of participatory schemes. Considering existing tools (ethical tools, legal frameworks within RRI and ethical review processes), the question of what would be a useful direction for new interaction modes/novel participation configurations remains to be specified according to the context (sector, scope and impacts, potentially affected publics, ability to address ethical issues and participatory options).

According to the context, the definition of the dialog configuration to be achieved can ensure deep-rooted participatory mechanism across ethical issues in R&I:

> One future challenge for public dialogue is to ask whether one-off dialogue events are sufficient or whether a more synoptic, far-reaching form of dialogue, deeply embedded in governance, is required – a key issue for nascent programs of responsible innovation. (Sykes and Macnaghten, 2013, p. 101)

A new participatory turn can be embraced, along with an assessment of the requirements of participatory mechanisms:

> It is also necessary to consider a number of practical issues. Cost is one such issue; another is the typical lack of incentives for lay participation in decision making. To the extent that practical incentives to encourage participation are missing, or that agencies consider the administrative costs unacceptable, then institutional development and experimentation are unlikely. (Fiorino, 1990, p. 239)

Another stream of challenges stands on the side of incentives and leverage that can support the democratic governance of publicly funded R&I with ethics and participatory mechanisms blended for the whole duration of the process.

The role of funders is crucial as "by administering research funds they are in a strong position to shape future research directions" (Stahl and Wakunuma, 2015, p. 25) and implement policy goals. Participatory mechanisms can provide added value for funders and can allow them to:

– enhance transparency and accountability to the public;

– bring knowledge that comes from working with a specific societal interest;

– improve links between cutting edge research and societal interests;

– improve the commercial viability of any innovative product that may be developed;

– give feedback during the research process (Stahl and Wakunuma, 2015, p. 25)[19].

Some general recommendations for the organizers of participatory mechanisms could be the following, especially in the case of funding structures:

– raise awareness of the issues to consider public engagement;

– allow public engagement to help shape the research agenda;

– create funding structures that are sensitive to public engagement needs;

– facilitate building connections between public engagement and researchers/innovators;

– emphasize the importance of dissemination and impact (follow-up and evaluation of project's results);

– celebrate positive R&I outcomes involving public engagement;

– ensure sensitivity to public engagement-related issues during evaluation[20].

Ethical public engagement in R&I also implies substantial funding support through targeted funding schemes. Surveys undertaken on this subject show major discrepancies across European Member States in the degree to which key public RFOs have developed an uptake of public engagement activities in funding schemes (Peter et al., 2018). Furthermore, public engagement elements are only marginally used as criteria in research proposal evaluations in Europe, with the exception of funding agencies of Nordic countries which use it to a large extent (Peter et al., 2018).

Furthermore, ethics and public–private partnerships can fuel publicly funded R&I with substantial support that addresses risk-taking: "there is a need to increase private investment, especially for innovations where the

---

19 These considerations on the inclusion of civil society engagement in funding calls can be used here with more general views on participation, not limited to civil society organizations (CSOs).

20 These recommendations have been adapted from the European project CONSIDER and specifically its recommendations for funders: CSO engagement has been here replaced by "public engagement" as a wider participatory category encompassing CSOs, citizens and stakeholders.

levels of uncertainty (technological, business model, regulatory, and user acceptance) are high" (Bogers et al., 2018, p. 10). Such partnerships can be confronted with the hindrance of demanding ethical and participatory schemes: the level of this inhibiting factor is yet to be demonstrated.

The core challenge might be the level of uncertainty and the way to confront it:

> In terms of impact, we also need to acknowledge that innovation does not always lead to results equally across organizations and people. In fact, while innovation can be the great leveler, it can also be the great divider. Uncertainty is also linked to the context we are living in and the trends we are experiencing. Specifically, there is uncertainty as to which emerging disruptive technologies we should publicly encourage in order to promote welfare. (Bogers et al., 2018, p. 8)

Participatory processes in R&I inevitably reinforce the ethical dimension as the closeness requires ethically sound decisions; however, participatory R&I always faces the risk of an unethical/interested use (Bergold and Thomas, 2012) if attention to the outcomes is not developed enough.

3.4.2. *Main critical factors facing institutions*

With the ever-expanding field of participation and deliberation, the organizations leading these initiatives will increasingly have to cope with legitimacy issues as long as ethical factors are not embedded in the participatory process in a comprehensive manner.

Could these practices lead to conclusions that are general and not sector-specific? How can uncertainty and unpredictability be addressed? Value-based and contextual categorizations of participation reveal the discrepancies between the various mechanisms termed "participation", as they have little in common in terms of their publics, outcomes, timelines and relationships with responsible innovation goals. In order to embrace the diversity of ethics perspectives, including the intersections of ethics and law, all approaches and RFO activity sectors were considered in this study, even though the discrepancies suggested above do not allow for a single definition of ethics or of participation.

	Critical factors
On ethics and law	Current regulations on the ethics of R&I do not cover the ethics of participation as such, as the link between ethics and participation is not developed enough. Further to regulations, ethical processes are already operative through formalized procedures in public funding schemes for R&I. However, ethics cannot be reduced to formalized/standard procedures only, and their occasional confusion with the legalization of ethics (soft law, ethical compliance) blurs the scope of their contribution. Also, ethics should not be considered as a toolkit but as a field (from applied ethics to meta-ethics; see Reber (2016)) that extends the regulatory schemes and helps decipher the legitimacy, the tensions and the adequacy of processes and legal compliance in regard to contextual criteria.
On the way participation is defined	There is no single approach toward participative innovation processes; however, the clear identification of activity dimensions, timelines, expected outcomes and types of participants can provide a common reference. In response to this need, the use of a common taxonomy and common indicators would be beneficial.
On ethical tensions	The ethics of participation provide orientation on priorities that are to be considered, especially in the event of emerging technologies and sectors of rapid expansion with uncertain societal impact/acceptance. The focus on such cases can provide insight on specific contexts and sectors in which participatory processes are especially decisive. Although some of the requirements to reach a high-quality level of participation can be convergent with ethics, ethics and participation are different. Indeed, ethics can focus on the types of relationships between actors, but may also turn toward problems to solve, where a more professional ethical expertise is expected in order to produce complete ethical arguments or counterarguments with a technical knowledge of the problems to assess.
On effective participation	A common reference base for the ethics of participation can help in setting criteria of good practices, as a combination of clear identification and matching of processes, actors and outcomes. Guidelines to reach effective forms of participatory practices would be of use both in top-down and bottom-up approaches, as common denominator and reference.
On the involvement of participants	Given the numerous ways to involve citizens or stakeholders in participatory processes for R&I, the questions of who is to be involved, by whom, when and for what also imply the clear identification of the publics: citizens and stakeholders are neither overlapping categories, nor are citizens (individuals) and organized civil society – they should be distinguished.
On a common framework for participation	Given the discrepancy across regulatory frameworks and ethical practices at the intersection of ethics and participatory processes, how could a common framework be established? As the main divide could be considered between the use of formalized/standard procedures on ethics of participation and the lack of formalized procedures and guidance, a common framework could be set in terms of criteria and understanding, on which an ethical approach to participation could be built.

Table 3.2. *Main critical factors to be considered in the design, implementation and assessment of participatory initiatives*

Before presenting a tentative ethics framework, some essential distinctions should be made as an ethical basis of any institutional attempt to design, implement or assess participatory practices. These factors are outlined in Table 3.2.

3.5. An ethics framework for participation

Following the assessment of actual practices through the scope of research funding organizations in Europe and the main findings of this case study, several needs and expectations have emerged as critical factors worth considering in the way "participation" is defined at the institutional level. Furthermore, although the scope of this case study was limited in terms of the surveyed RFOs, the conclusions reached highlight some difficulties, previously mentioned in the theoretical analysis based on the scientific and gray literature. The need for stabilized taxonomies appears as the pivotal element and pillar in future endeavors aiming at developing ethical participatory practices. More fundamentally, the need for a context-sensitive approach and the capacity to alternate between stable considerations of key factors and an adapted consideration of the contextual (and sectoral) challenges (and possibilities) seems inevitably bound to the success of such an effort.

Further to the previous empirical and theoretical considerations, some main conclusions can be considered for the shaping of future ethics frameworks for participation by institutions. This last section successively presents some general indications summarizing the main highlights of the analysis, as well as an attempt at an ethics framework.

3.5.1. *Some guidelines for future ethics frameworks*

Summarizing some essential indications that could be worth considering by R&I institutions, the following findings might appear as the first basis for the definition of future guidelines.

Table 3.3 summarizes the main factors that would allow for the design of an ethics framework for the effective, fair, legitimate and responsible design/implementation/assessment of participatory practices in R&I. Despite the difficulties in adopting a stable instrument to ethically frame participation, such considerations could arguably ease the process on behalf of R&I governance and reflect more consideration for the participants involved. Provided key factors are considered, the added value of participation would be established more easily, with a clearer understanding both for R&I governance and the participants, and the possibility of referring to concrete factors. In that sense, these key factors are a concrete outline of the ethical boundaries, without which participatory processes remain in a hazy area.

	Dimension	Sensitive factors faced by R&I institutions
ETHICS	Main ethical challenges	– In procedures: difficulties in the implementation of evaluation procedures (adaptation of ethics reviews for emerging technologies and new ethics issues; lack of national ethics committee structure; unequal portfolio of reviewers across countries with issues of conflict of interest or diverse evaluation exigencies) – At the conceptual level: improvement of procedures (clarification of what is funded, responsibility, accountability, transparency); diverse understandings of ethics – Awareness raising: identification of red lines for sound procedures in public funding of R&I; legitimacy; public dialog on controversial issues; addressing ethics of R&I at civic/societal level
	Ethical challenges in participation regarding: i) projects ii) evaluation iii) participants	– Interconnectivity and equal importance of the three dimensions – In (i) funded innovation projects: difficulties in identifying benefits for participants and in structuring participation in an ethical way; identification of the type of responsibility; unequal development of R&I sectors in funding programs (procedures); boundaries of public investment in regard to industrial developments – In (ii) projects' evaluation: discrepancies across sectors (lack of recognition and visibility across sectors; gap between scientific and ethics evaluation in some cases) and across countries (different evaluation models or lack of ethics evaluation) – In (iii) new actors' involvement: challenges that encompass ethical challenges related to the funding structure and to evaluation practices; unclear landscape, without guidelines; the issue of the quality of involvement; the issue of effective inclusion of participation in the R&I process; the issue of biases and imbalances

	Dimension	Sensitive factors faced by R&I institutions
PROCEDURES	Priorities for participants and activities	– Context-related priority setting according to the purpose of each R&I process – Importance of representativity – Quality of participation and dialog – Clarification of types of participants: inclusion of all voices; focus on citizens as members of civic society; distinction of individuals from organized civil society; diversity of actors; involvement of neglected groups as local authorities and trade unions of salaried employees
	Selection of participants	– Context-specific selection guided by dialog to identify adequate participants – Representativity of civil society, avoiding narrow or "comfortable" selections, with criteria of diversity, relevance and balance – Diverging views on who should select participants: i) top-down selection with recruiting procedures led by experts, following above mentioned criteria; ii) bottom-up selection (citizen-led)
	Sensitive contexts/sectors in participatory processes	– Specific contexts/sectors particularly sensitive to participation that should be prioritized: gene technologies, energy, environment, health, AI – The methods and structure of participation prevailing (importance of power relations; responsibility; transparency; accountability issues) – The interpretation/understanding of contexts prevailing (quality of expertise and understanding of expectations; benefits; consideration of social context)
	Priorities for participation in RFO processes and in projects supported by RFOs	– Equal importance/priority and complementarity of efforts in both dimensions – Need for a methodology of participatory instruments in core RFO activities; need for a wider adoption of ethics reviews; need for better guidelines to implement policy; priority on RFO programs focusing on public engagement or citizen science and also improvement in the implementation of RFO structures (e.g. transparency), reflecting societal dynamics – RFO activities already have ethical compliance procedures, but there is no guidance on projects supported by RFOs which could be prioritized in the development of participatory practices (also, forms of open innovation could gain some guidance as well, up to a certain extent)

	Dimension	Sensitive factors faced by R&I institutions
OUTPUT	Added value and justification of participation	– To be found in the quality and value of procedures (broader and deeper understanding; more fair and legitimate processes; broadly shared benefits; deeper reflection; sustainability of decisions) – In a better connection with societal needs and social reality (inclusion of grand societal challenges; diversity of data and perspectives) – In the better functionality of R&I processes (participation brings better acceptance and integration and better value of public investment in R&I)
	Goals and outcomes	– Value-oriented goals and outcomes (democratization, openness, acceptance, transparency and legitimacy of R&I; deliberation; full participation of citizens in all R&I stages) – Through efficiency indicators (increase of quality, relevance, acceptability and applicability of R&I outcomes, with real impact on innovation and policy; economic efficiency) – Mainly external (to the R&I processes) or unspecified outcomes that cannot be settled beforehand: for example recognition of participants
	Quality criteria of participation	– In values: ethical participation (social justice and equality in citizen empowerment and representation); regular and deep engagement with control mechanisms in decision-making; creation of enabling environments for a diversity of perspectives – In processes: quality of dialog in participation; financial and structural support for dialog as core activity in R&I processes; better relationship between R&I actors and participants – In results: measurement of outcomes at each stage through indicators; better framing of participation; focus on outcomes and usefulness of R&I results and social impact

Table 3.3. *Factors worth considering in future ethics frameworks for participatory practices*

3.5.2. *A tentative ethics framework*

Further to these recommendations, an attempt to design an ethics framework was undertaken, encompassing all the theoretical and empirical findings of this study. The ethics framework[21] was designed and tested in the course of the PRO-Ethics project, aiming at new solutions to support the design, implementation and evaluation of participatory practices in publicly funded R&I at the European level.

This tentative ethics framework offers a set of reflections, tools and criteria to assess the ethical aspects of novel interaction modes in R&I activities. It is structured around a set of questions and associated actions to consider when designing, implementing and evaluating a participatory process. The aims of this ethics framework are: to provide orientation and legitimacy for participation, as well as addressing some practical issues in implementing participation; to ensure participatory processes are meaningful both for participants and organizations; and to certify that the design and implementation of participatory processes considers all main ethical aspects that need to be addressed, prior to, during and after its completion. As such, it operates as a tool for reflection and justification and a reminder of the main ethical aspects to consider in a comprehensive manner, both in terms of timeline and of factors to consider.

This framework aims at ensuring citizens and stakeholders' participation can be properly designed and implemented, addressing different contexts, resources and needs. The ways participation is conducted, how it is based on legal and regulatory frameworks and to what extent ethical issues are considered can differ significantly across countries and contexts. At the same time, concerns over non-traditional stakeholders (such as citizens or NGOs) and their potential added value emerge along with new modes of innovation and open R&I configurations. In this respect, PRO-Ethics develops an ethics framework with principles, guidelines and tools that can provide some guidance and a roadmap to ensure citizens and stakeholders' participation is properly put in place without disregarding principles of fairness, transparency, gender equality, privacy and sustainability.

As participation requires a significant amount of resources (both human and financial), expertise and commitment, the need for guidance and tools to design and implement processes that are both adaptive and legitimate is an issue that needs to be addressed. In order to do so, the underlying guiding question of the added value of participation has to be addressed, while notions of accountability and transparency toward participants shape the

21 The tentative ethics framework can be found in Appendix 2.

boundaries of the present framework. Taking a step backward to analyze the grounds on which participatory processes can be ethically developed and conducted appears to be a requirement that connects with the need for guidance and indicators.

Each context having its own specificities, this ethics framework only provides orientations and opportunities to consider, setting some common denominators aiming at covering all the main ethical considerations and ethical requirements addressed by any participatory process in R&I. The tools and guidelines provided consist of: (1) taxonomies and definitions and (2) a list of questions with related "actions" to be undertaken. The taxonomies relate to participation actors and participation activities, in order to specify what or who is targeted under the broad notions of "citizens" or "stakeholders" and "participation"[22]. In addition, a list of questions to be addressed before, during and after the implementation of participatory processes aims to ensure all the main ethical aspects are adequately addressed, from the design to the *ex post* assessment. These questions cover: the design of the participatory process; the way to address the classifications provided; the main ethics issues and risks and how to properly identify and address them; the anticipation of critical factors that may arise in specific contexts; the monitoring and evaluation of participatory processes; and the identification of the added value of participation both prior to and after the implementation.

The comprehensive timeline of this framework was designed in such a way as to integrate the different chronological dimensions of participatory practices: the framework is to be used for the design of participatory processes, during the implementation and after the implementation, in order to ensure that some of the main ethical dimensions are addressed at each stage. This framework can be used as a reflection and assessment tool, as well as a roadmap helping to keep track of the decisions taken in a consistent manner. In this regard, it highlights some key questions that can help decide when ethical expertise is insufficient and should be externally sought, as well as which participatory features should be considered and how, and the reasons why a participatory initiative is considered in the first place and the way it relates to the wider societal context.

---

22 These categories and their definitions are those that were presented at the end of the first chapter of this book.

In this respect, this ethics framework consists of a common reflection base that provides guidance for the proper adaptation of participatory processes. It also allows for flexibility and enables reflection on the reasons why participation is considered in the first place and if it is an appropriate tool to use in a given situation – as its added value is not taken for granted.

Conclusion

The notion of "participation" in R&I can refer to a variety of practices and means that connect differently to ethical issues. The challenge of defining participation is all the more difficult as interferences between various dimensions both in theory and practice blur the landscape of what is termed "participation" in the field of R&I. The common thread throughout this study has been analyzing this stratified field in a way that could provide some general orientations, in an attempt to classify and explore the various levels – not exhaustively, but methodically. By setting the bootstrap of this critical review in the opening-up of all the meanings of "participation", the varying approaches of participatory practices have shown the extent of a field that encompasses a great variety of features, which do not always share common ground, neither in processes nor in actors and outcomes. The diversity of meanings suggests the need to consider the root causes behind participation and the expected outcomes; the nature and scope of the ethical issues; who is affected, interested or can contribute to solutions; the timescale; and the resources.

Overall, participation might be considered as the best resource to enable non-conventional views and thus help in tackling the factor of uncertainty through shared responsibility. With the objective of continuing to enhance the science–society relationship in response to new trends, the European Commission has taken this commitment a step further thanks to RRI, which has highlighted the essential contribution of participation to the ethics of R&I. RRI has opened a space for experimentation in R&I participatory practices that keeps evolving, alongside innovation issues and societal challenges. Under the umbrella of RRI, a strong European commitment to the promotion of sustainable and ethically sound R&I throughout its various stages was made. Across formalized procedures (ethics reviews) and a wide set of policies that nurture research ethics, research integrity and social/societal implications, policy-making has provided a wide spectrum of ethical standards, norms and regulations. For publicly funded R&I, ethical compliance has allowed for the monitoring of interests and ethical conflicts and their better identification, which could be considered as an asset in terms of the qualitative guarantees it allows. The literature on RRI has highlighted this extension of the science–society discourse through public engagement and the benefits of the co-production, upstream engagement and reflexive responsibility of science and innovation outcomes. Addressing the loss of public trust for science and innovation advancements, participation intervenes as a remedy, although a restrictive approach may restrain its potential to mere assessment schemes.

By detecting regulatory gaps and combining ethical issues with participation (broadly defined), new indicators of efficiency arise. This study has shown the precedence of ethics for laws and regulations. By putting forward the notion of responsibility, RRI has opened new avenues of reflection. However, theoretical considerations rely on the capacity of public schemes to be attractive without falling into the trap of over-regulation and overly restrictive frameworks. In the path of responsibility in innovation, the way forward calls into question whether the existing operating governance systems for R&I – standards, norms and regulations – are adequate for achieving desirable social outcomes. Ethical participation bears a variety of requisites stemming from responsible innovation and ultimately ensures that innovation processes are responsive to the needs, values and expectations of society (Van de Poel and Sand, 2021). In this sense, R&I that is responsive to society ensures that we engage in "pathways responsive in the face of uncertainty" (Owen et al., 2012, p. 758). The benefits of publicly funded R&I might include the possibility it offers for renewed questioning of ethics, benefitting from their prevalence over regulatory schemes.

As regulations are particularly limited, this study combined legal resources and soft law with scientific literature in an attempt to bridge the gaps. The ethics part is regulated via the GDPR for data protection, some complementary international codes on research integrity (mostly concerned with research on human participants) and some voluntary codes to which some European universities have committed which are not legally binding (at the national or European level)[1]. Regarding participation, only a few conventions and rules have been codified into laws with a degree of binding legal status. The Aarhus Convention is probably the most prominent one, applicable in the context of environmental regulations, but with particular relevance for the context of R&I ethics too.

Ethics provide adequate tools for an expansion of existing policies and regulations – the procedural dimension of compliance – towards the identification of participation as the means to implement democracy in R&I processes, embracing voices that may be at odds and understanding conflicting values or principles. Deliberative democracy enshrines participation as part of decision-making and, although the value of deliberative systems can vary, the need for clarification about the understanding of participation is a common requisite. Also, as most scientific resources insist on early engagement as well as the importance of social impact evaluation, the benefits of a long-term participatory approach override fixed-term processes. As the *in itenere* assessment ensures an ethically sound evolution throughout all R&I phases, similarly, other dimensions of participation seem to gain from the temporal consideration of R&I, thus allowing for a satisfactory level of outcomes.

On a theoretical level, scientific and gray literature provide guidance on criteria, indicators and more generally options helping to classify the internal diversity of the subject of this research. On an empirical level, the regulatory practices in R&I do not provide information about participatory practices, despite their advocacy for the benefits of citizen engagement or what is commonly termed "public participation", embracing all features. The complexity of participation is also due to the diversity of ethical grounds supporting participation, which can vary according to the normative/moral justifications that are considered, and which can either refer to ethics (ethical concerns, ethics broadly understood), policy (ethical reviews, RRI) or laws (regulations). Combining theory and practice can only result in a predominance of theory as guidance for the multi-faceted landscape of practices, across countries, regulations and policies.

1 The European Union has issued ethics guidelines that come as a complement to these legal resources.

Selecting the "right" participatory process and ensuring its proper implementation are two aspects that are also dependent on the way the value of participation is perceived and justified. Participation is not always sufficiently recognized as an added value, while sometimes there is a gap between an arm's length consultation with genuine participation by all parties and the subsequent inclusion of views in the decision-making process. The selection of participants is another element in need of guidance and improvement, prompting the idea of the creation of a pool of experts to supervise and help participatory processes. According to the specifics of each case in terms of scale and purpose of participation, selection of participants can be random if only general targets are expected; however, in most cases, a more applied selection could be useful in order to reach qualitative input. Also, differentiating individuals (as citizens or end-users) from civil society organizations is another important element worth considering.

The proper identification and selection of the right participants, the adaptation of the participatory method to the values of research integrity and the ways to deal with different approaches are part of the way forward. Tackling the diversity of organizational contexts, R&I institutions can benefit from these guidelines and methodological guidance on ethics and on participatory processes. As the evolution might be in conflict with existing funding structures and processes, future steps may hinder comfortable selections and procedures and lead to methodological gaps. A solution could be to have an instrument to deal with this diversity of views, in order to enable all RFOs to decide what kind of participation is needed and for what, in each specific case: currently, only a few European countries have RFOs with effective competence in adaptation. In these future evolutions, some of the challenges will be to properly import in RRI all quality criteria for participation that could be meaningful across all countries and sectors, to define the purpose of participation and its means and to make the distinction between ethics and legal compliance, leaving space for ethical interpretation and reflection. This comprehensive presentation of existing definitions, possibilities and indicators clearly indicates that participatory practices might be considered as a very promising component of innovation ethics, provided their understanding and configuration are framed at the outset.

Formalized participatory models are facing the tripartite distinction between information, consultation and participation as such (in decision-making), a distinction that could be considered as the common denominator across participatory processes. Similarly, the distinction between participants defines the related diversity of their contributions, depending on their roles as citizens/lay people, stakeholders, experts, politicians and policymakers. The objectives of participatory processes are also another dimension with internal diversity, since active contribution to decision-making differs from advising, whereas the output of participation can diverge depending on whether it is oriented toward diversity or consensus. Considering all these parameters entails a more precise adequacy for participatory practices according to the context and reaching "active" forms of participation.

With the leverage of public funding, the embeddedness of participation in the ethics of R&I can act as a powerful multiplier. In the absence of a straightforward connection between ethics and participation, some good practices can be underlined in the direction of the optimal inclusion of challenges raised on both sides. Also, the promotion of the ongoing process of participation throughout the whole timescale of research and innovation, from inception and design up to outcomes, can ensure both public acceptance and process quality regarding the (final) social impact.

The focus on RFOs among other institutions was intended as a specific case study that could help to define the different participatory practices in R&I, and thereby to reach a more granular approach than the general label of "participation". RFOs can be considered as particularly indicative of the way participatory practices are handled at the governance level. The potentially high leverage of RFOs regarding the application of frameworks, combined with their expertise in the design and follow-up of R&I programs, endows them with valuable experience worth considering in this analytical endeavor. However, the main issues remain unresolved in institutionalized participatory practices in Europe, where the connection with ethics is highly dependent on the compulsory checks and balances encapsulated in European funding schemes.

Despite some common denominators, there are major discrepancies in Europe at the level of ethical values promoted, as well as in RFO procedures, means and organization, but also in sectoral specificities urging for adapted guidelines and in the selection and implementation of adequate participatory processes. The current structure of RFOs is in most cases expert-led, since participatory processes usually involve experts, research institutions and scientists – across all RFO activities. In addition to expert-led participatory R&I, citizens and end-users are generally poorly involved in RFO activities, both in current and in expected future activities. As an exception to this, projects' evaluation and social impact measurement are two types of RFO activities that are more receptive to less represented types of participants (citizens and end-users). This is, however, tempered by the relatively low proportion of evaluation and social impact activities among the declared current and future activities of RFOs.

At the level of ethics, the main highlights are that R&I ethics is mostly represented in dialog with arguments – deliberation – and that regulations alone cannot ensure any ethical progression. From an ethical point of view, the way ahead is not one of relying on regulations. There seems to be a common understanding that regulations alone cannot ensure any ethical progression. As ethics is ahead of participatory processes, there is skepticism about what can be solved with ethics, and standard formalized ethical thinking may not always be of help in RFOs' processes and practical outcomes (funding). The need for a more ethical approach implies focusing on better relationships and conditions for dialog among different actors. Moreover, as the ethics of R&I do not always differentiate research from innovation (with their subtle differences), it has to be taken into account that these two represent different normative systems.

In light of these needs and the importance of clear distinctions regarding categories of participatory modes and participatory actors, the contextual justification and the expected added value, a comprehensive approach to the ethics of R&I and participatory practices seems of utmost importance. In the attempt to encapsulate all these needs for guidance, a methodological ethics framework was outlined, as an open and flexible toolbox. Considering that a single approach would not be possible given that the determining contextual factors only allow for case-by-case guidance, the proposed ethics framework in this book is a guide for addressing the questions that are worth considering when designing, implementing or evaluating a participatory process in any field of R&I. To embrace the sectoral diversity of R&I and the different timelines, this common guide specifies a list of actions and related questions to be asked, at each stage of the process.

These findings highlight the need for R&I institutions to develop unified guidelines and implement them: this could prove difficult, given the differences in structural means, understandings and aspirations. In the case of evaluation procedures, if the same standard of ethics review is adopted across countries and sectors, it would be timesaving for basic ethical dilemmas – thus avoiding unnecessary discussion that hinders action. As a minimal ethics procedure, it cannot, however, overshadow the ethical deliberation, the monitoring of projects and the decision about a project's eligibility to be funded. Ethical evaluation can be useful in helping a project team if the project is scientifically sound and deserves to be funded, but denying funding because of ethical reasons without further examination could be counterproductive. Also, enhanced ethics procedures are needed in some specific sectors, such as the medical sector, where ethical deliberation can intervene on top of an ethics review but where standard ethics evaluation procedures ensure a standard of care and shared responsibility. Ethics reviews could be perceived as a baseline, ensuring the basics – or non-negotiable parts – and that principles are covered, prior to considering deeper insights. As such, ethics reviews could be considered as the main tool ensuring research integrity as a component of R&I quality, preventing harm such as malpractice related to research or data management, for instance: as such, guidelines going in this direction could be uniformly compulsory. National adaptations following general principles could intervene to ensure evaluations are context-relevant; however, guidelines for ethics reviews, the training of experts or the distinction of good practices could prove particularly efficient at the European level.

The very problematic nature of "participation" raises the question, at the governance level, of the possibility of developing a unified framework. The configuration of the involvement, the timescale and the final outcomes of the processes can entail either strong or poor forms of participation. Depending on the stages of innovation processes, public participation – generally considered – embraces a variety of actors, who can either be lay people, persons directly involved in the R&I process, experts, individuals or groups representing specific interests. These audiences can relate either by interest or as affected publics, whereas future societal implications override the possibility of clear identification of participation actors and the configuration of their possible involvement. The closely woven field of R&I reveals the lack of clear identification of participatory practices within R&I ethics. Consequently, the ethics of participation appear as a new playground for experimentation, which can draw from general considerations from the scientific literature in the well-developed field of deliberative democracy, notably, or in institutionalized participatory forms such as citizen science, which allow for more precise identification of processes and outcomes.

Following the findings of the case study and of the theoretical analysis, this study suggested the design of an ethics framework for participatory practices. Some main critical factors were listed as the basis for any future governance initiative in this direction. Further to this, a tentative ethics framework was proposed, bringing a clear outline of distinct categories and implications of "participation", and which can be used as a toolbox to test the various levels, depending on where, how and with whom participation is considered. Criteria can help in the identification of good practices, as contextual elements can be considered and connected to the configurations of participatory processes. At the same time, new digital technologies can reconfigure democracy and reshape the field of participatory practices by relocating the various components of participation.

Ethical practices of R&I institutions are context-sensitive, similarly to their participatory practices. Having a comprehensive overview at a European scale would imply considering national specificities pertaining to the financial, technical and human resources needed to frame and support publicly funded R&I. National and sectoral discrepancies have been brought to light on both sides, in RFOs' current practices and intentions of future development, as well as among the viewpoints of the experts interviewed. In the absence of unified practices or tendencies in evaluation procedures and in participatory processes – let alone the variety of approaches to ethics – it seems practically impossible to consider ethical practices for public funding structures at a more general level.

While it seems important to avoid over-regulation and to consider separately specific concerns in some sectors, the backbone of a comprehensive framework could be a combination of general guidelines regarding ethics and the usual dilemmas (such as in accountability or decision-making) and national or sectoral specificities that should be considered to ensure the efficiency of guidelines across all contexts. Although no context or sector should prevail, a case-specific approach could complement some main normative principles indicating how to maneuver in each specific case, without contradicting sectoral boundaries but allowing for a responsive adaptation of citizen and stakeholder participation according to the R&I process and society's needs, norms and expectations.

Appendices

Appendix 1

Categories Considered in the Case Study on Research Funding Organizations

The following lists offer a detailed view of the categories considered, as well as their definitions, in the case study on RFOs. These tentative categories and definitions served as the first canvas for the categories and definitions proposed in the first chapter of this book and the proposed ethics framework. As such, these categories and their definition have been a work in progress and should only be considered to understand the results of the RFO survey and the related diagrams. In terms of stabilized definitions for participants and participatory activities, those included in Appendix 2 – and previously presented in Chapter 1 – are the upgraded version of these ones and should be considered as the final proposition.

This study, part of the PRO-Ethics project, served to reveal some interesting conclusions in the practices of RFOs at the European level. Although the scope of the study was limited, the use of precise categories and definitions served to underline some tendencies and counter-intuitive situations, which have been emphasized through the use of the following specific categorizations. These categorizations differ from the ones suggested in section 3.4 of Chapter 3 since they were outlined at an earlier stage, thereby serving to identify the coherence of categorizations in participation actors and participatory activities.

For the purpose of the RFO survey, 10 main categories were outlined and suggested as options for participants. Although some categories partially overlap, this distinction was helpful in avoiding unclear answers as well as in measuring the degree of precision that should be maintained in future taxonomies. The categories of participants that were outlined were the following entities:

– *citizens*: this category includes the general public, lay people and citizens as persons (or collectives) with civic expectations[1]. Moreover, since sometimes end-users are also citizens, this distinction serves to underline the general dimension of involvement, referring to the broader sense of "public participation";

– *end-users/consumers*: oftens close to "citizens", end-users/consumers are defined as a specific category among citizens, referring to beneficiaries of the end product of the R&I process. This category may also relate to a broad participatory approach that still remains more specifically oriented toward the impact and the uses of the products of R&I activities in society;

– *experts*: this category serves to identify individuals enrolled as experts in RFO activities, but at the individual level and not as part of any of the other categories. Therefore, lay experts can be included, for instance, if their involvement mainly values their expertise and if this category prevails over others (e.g. "citizens"). Experts can be individuals with sectoral expertise, such as for example ethics experts who are part of the ethics review processes (proposals/projects are assessed by ethics experts, who identify the main ethics issues related to the implementation of a project);

– *firms*: this category includes large companies, as well as small- and medium-sized enterprises (SMEs), considered as beneficiaries (actors who implement the funded proposals) or non-beneficiaries of funding schemes and involved by any means in RFO activities as participants;

– *governmental institutions*: also understood as public bodies, this category encompasses national public bodies (e.g. government departments or ministries) that might be involved in the monitoring of RFO activities and projects. They relate to ethics through ethical compliance mainly, ensuring ethics requirements are met in grant agreements signed with the beneficiaries at a national level;

– *innovators/entrepreneurs*: this category includes individuals contributing to R&I activities, not as representatives of a firm or an organization, but

1 In this taxonomy, the main boundaries for the "citizens" category are anchored in the citizens/stakeholders distinction.

selected individually for their specific experience in entrepreneurship and innovation. This category also comprises social entrepreneurs;

– *NGOs*: NGOs are civil society organizations, which have different knowledge and leverage compared to citizens;

– *research institutions*: this category comprises researchers involved in the implementation of projects (mainly) on behalf of their institution. Researchers are meant as physical persons working for a legal entity (SMEs, universities, research institutes) conducting research and to whom public or private funding has been granted to do so;

– *research participants*: this category refers to human participants who are part of the subject of research (for instance, in medical sciences, public and patients' involvement);

– *scientists*: this category includes individual scientists involved in RFO processes not on behalf of a research institution but for their own scientific knowledge[2].

This detailed categorization proved useful throughout the study to ensure a differentiation of individuals from institutions whenever this was meaningful.

Similarly to this taxonomy for potential actors that can be labeled as "participants" in RFO activities, some types of activities were identified in order to obtain a more specific picture of the type of each process considered in the larger scope of R&I processes. The main types of RFO activities where participatory processes can be implemented may refer to – broadly speaking – the program level, the project level or the RFO-wide level (such as policies). For the purpose of this survey, seven types of activities were considered:

– *communication* of funding activities: one-way engagement toward any type of public, sharing information related to RFO (funding) activities;

– *consultation*: processes of engagement with any group of citizens or stakeholders, in order to obtain feedback on policies and programs;

2 Elaborate categories such as "scientists" and "innovators/entrepreneurs" have been included in this taxonomy in order to separate cases where participants are involved primarily under this identity rather than as "experts", for instance, hence being counted in the prevailing category.

– *evaluation* (of projects): the ethical and scientific evaluation of project proposals in funding schemes, as well as the interim and *ex post* evaluation for projects that receive funding;

– *knowledge sharing*: dissemination of information on R&I contents, outputs and scientific knowledge. Also one-way like the "communication of funding activities", knowledge sharing relates to the dissemination of scientific content only;

– *monitoring*: technical monitoring of funded projects in the context of RFO funding schemes[3];

– *program design*: the identification of program objectives and of R&I priorities, resulting in the definition of programs that will be submitted for funding opportunities;

– *social impact*: RFO activities that aim at measuring the social impact of R&I processes, including the development of indicators and the monitoring allowing such measurement.

The surveyed RFOs operate with different and sometimes only partially overlapping working definitions of participants and participation modes, which made direct comparison difficult. The above taxonomies were developed in dialog with RFOs to alleviate this issue and enable the future tracking and documenting of participatory processes in a comparable way.

3 Usually an activity carried out internally, although *ex post* monitoring of results can involve other actors, in addition to the involvement (feedback) of programs' beneficiaries.

Appendix 2

Tentative Ethics Framework

How to use this tentative ethics framework

The following tentative ethics framework has benefitted from real-life testing in the course of the European project PRO-Ethics. Being a draft ethics framework, it is presented in this book as a suggestion of what an ethics framework could look like.

The present ethics framework consists of principles, guidelines, assessment criteria, good practices and proposals relating to regulatory environments, with a focus on how citizens' engagement can be properly put in place without disregarding ethical principles of fairness, transparency, gender equality, privacy and sustainability. With the ambition of addressing the ethics of participation in the context of publicly funded R&I, this ethics framework aims to operate as a common standard on how to deal with a diversity of views and how to properly address ethical issues and ethical risks to be used both in top-down and bottom-up approaches.

In this perspective, it is intended to operate as a common reference on the ethics of participation for policymakers, funders and civil society. Taking stock of the theoretical difficulties that surround the lack of a joint framing of ethics and participation in R&I, this framework aims to provide information on how to deal with the diversity of views on ethics and participatory processes and the diversity of practices on behalf of RFOs, both in terms of legal and structural means.

This framework is meant to operate as a *roadmap* for all RFOs, highlighting some important ethical features that are not context-specific. In that respect, the framework is intended to cover all stages of participatory activities (design, implementation, feedback) applying to either design, development or output phases of R&I processes. It also provides a list of questions and operational steps to consider for each ethical dimension and some clarity both with regard to the design and the implementation of participatory processes, as well as social impact evaluation. Thanks to its format and scope, this framework can respond to any specific context of implementation, providing a list of key topics and questions to be addressed in order to *determine whether an ethical participation process is possible or necessary, and if so, how it should be structured in view of the different factors that should be integrated.*

This ethics framework is structured in two parts:

1) A general descriptive part (theoretical introduction) explaining the context of use and presenting the scope, the objectives and the framing of this ethics framework 0.1 and how it is meant to be used.

2) A second part, of an operational nature, providing the tools and guidelines, and also a general glossary, and presenting the main categories (taxonomies) and definitions (which are essential components of the ethics framework), as well as broader definitions of frequently used terms. The second part – relating to tools and guidelines – of this framework has the format of a table in order to be as directly operational as possible. Divided into separate sections, these recommendations and guidelines take the form of "actions" to be implemented, in a similar way to a checklist, addressing all main ethical considerations. As such, *this list of questions should be asked prior, during and after designing and implementing a participatory process*. It aims to cover the main ethical aspects – regardless of the specificities of each context – as a list of actions and verifications to be undertaken that can ensure an ethical participatory process. This roadmap is meant to be understandable and meaningful for a broad audience, beyond Europe and beyond RFOs.

Defining a methodology of participation is a process that needs to consider diverse perspectives and a combination of clear identification and matching of processes, actors and outcomes. This framework can provide guidance both for innovation projects and innovation funding practices, thus addressing participatory R&I practices supported and implemented by regional and national RFOs. In terms of criteria and understanding, this ethics framework also takes into account the current discrepancies regarding the use of formalized/standard procedures for ethics of participation, as there is a lack of guidance and formalized procedures relating to ethics and participation in some national contexts. Citizen and (non-traditional) stakeholder participation can take many forms and a variety of subtle distinctions have to be embraced in order to *specify the exact conditions under which an individual or a group is or should be invited to participate, with what tools, knowledge and expectations, when, how and with what outcomes and impact on decision-making.*

Box A 2.1. *Note on the use of the tentative ethics framework*

Part 1 – General considerations

General considerations on ethics

Although some of the requirements to reach a high-quality level of participation can be covered by ethical compliance, the consideration of ethical grounds may result in ethical tensions (legitimacy conflicts between ethics and participation). Indeed, ethics can focus on the types of relationships between actors but may also turn toward problems to solve, where a more professional ethical expertise is expected in order to produce complete ethical arguments or counterarguments with a technical knowledge of the problems assessed. However, the design of ethical processes addresses the issue of the right momentum, as ethical guidance may be ineffective once technology or R&I processes are already developed, thus urging anticipation of future consequences through ethical discussions when they can still be used to inform the R&I processes and decisions (Van den Hoven et al., 2014).

Considering the *complexity and unpredictability of innovation environments*, the concept of responsibility could be considered as a key dimension in R&I. This ethical stance also serves to prevent R&I processes that could be considered “irresponsible”, mainly occurring in contexts where the importance of an innovation’s societal context has been miscalculated or when unresolved conflicts have taken place during the innovation process (e.g. lack of precautionary measures or neglect of ethical principles). The concept of RRI has emerged as an extension of the Science in Society discourse about the co-production of solutions to global challenges and purposeful science, upstream engagement (Wilsdon and Willis, 2004) and the reflexive responsibility of scientists and innovators (Owen et al., 2012). For this perspective, ethics cannot be reduced to formalized/standard procedures only, and their occasional confusion with the legalization of ethics (soft law, ethical compliance) blurs the scope of their contribution. Also, ethics should not be considered as a toolkit but as a field (from applied ethics through to meta-ethics (Reber, 2016)) and a specific perspective that extends the regulatory schemes: *ethics helps decipher the legitimacy, the tensions and the adequacy of processes and legal compliance in regard to contextual criteria.*

The *pluralism of moral theories* suggests that, while deontology (assessment based on universal principles and duties) may seem to be the most relevant approach, there are several rational paths that can be followed through ethical pluralism in regard to ethics assessment (in justification context): types of entities assessed (i.e. action, person, institution and technology) from a normative ethics perspective; normative factors (i.e. values, consequences, virtues or norms); and foundational normative theories (ways to select normative factors and types of entities). Conflicting factors or hybrid forms of reasoning indicate the need to move beyond regulations (as in the ethics review/assessment) and embrace a broader, pluralistic, scope. These views imply an enhancement of reflexivity and responsibility, this latter being polysemic and subject to different understandings.

Ethical assessment procedures and the ethics review

Given the *precedence of ethics over positive law and regulations*, ethical compliance and ethical appraisal (assessment) procedures such as the ethics review do not cover ethics but are rather closer to legal standards and regulations. The current ecosystem of publicly funded R&I has an advanced set of ethics assessment procedures, ensuring the compliance of funded research activities with fundamental ethical principles. However, the ethics review process differs across countries, since such formalized ethics procedures usually systematically occur in the context of funding programs supported by the European Commission. The ethics review[1] has a different impact on projects depending on consortiums' structure and sector of innovation and the involvement of non-EU countries. The current European landscape of regulatory bodies reveals gaps and difficulties in the merging of ethical practices with participatory practices in R&I. Overall, the study of resources across ethics regulatory bodies shows that the connection between

1 The ethics review process takes place in the assessment of R&I project proposals within European funding schemes. This procedure has been established since 2011 and involves proposals that have been identified as raising ethical issues being further examined by a panel of experts.

ethical practices and participation is not yet developed enough, as the connection is often unspecified, maintained at the level of a general appreciation of the potential benefits in keeping a transparent relationship with society across R&I processes, mainly in regard to matters of public acceptance and public trust.

On a wider scale, the consideration and practice of ethics in R&I requires specific skills and knowledge that researchers, scientists or innovators usually do not have. Ethical analysis encompasses identifying the ethical aspects of a project or issue (the ethical questions or moral dilemmas at stake); familiarity and conformity with ethical standards; and the adddressing of ethical dilemmas when moral principles and specific methods to build, recognize and justify ethical issues are in conflict. The evaluation of right or wrong is based on ethical/moral values (ideals), or principles and norms that define standards, which can be identified as "ethical principles" on a general level, some of them being principles concerning individual rights, principles concerning benefits and harms, fairness principles and virtues. By bringing a proper identification of ethical issues, ethics expertise provides guidance for R&I processes, a reflection on the consequences of the choices made, an enhancement of transparency (to a certain extent) and accountability as well as better quality in processes. In this respect, *ethical expertise provides tools to address the complexity and unpredictability of innovation environments, ensuring responsibility within R&I processes.*

The concept of responsibility can either refer to the legal, moral or social sphere. Sometimes considered as a term for responsibility, the precautionary principle[2] is fundamentally of an anticipatory nature and "does not allow uncertainty on the scientific side of assessment to be used as an excuse when serious presumptions of significant and/or irreversible damages have been made" (Reber, 2018, p. 40). The diversity of cases highlights the importance of an assessment of the state of affairs in science and the type of uncertainties involved. Indeed, "uncertainty" can relate to several cases: dealing with hypothetical effects and imaginary risks that require independent scientific evaluation, both transparent and publicly accessible;

2 The precautionary principle was first embedded in European policy in the 1992 Maastricht Treaty for environmental policy; it has expanded to other fields of policy under EU law, where it intervenes as provisional risk management measures that have to be considered to prevent adverse effects. See the *PRO-Ethics Paper Manuscript on Participatory Practices and Ethics Issues in Innovation* (D1.2).

facing a defined and quantified risk, which implies that policymakers can respond with a normal risk management approach; situations in which one cannot fully rely on the scientific information system as such when it comes to the estimation of possible adverse effects requiring a precautionary approach; and when particular cause–effect relationships cannot be scientifically established, although adverse effects are known (Von Schomberg, 2012a, pp. 151–152).

General considerations on participatory practices

Participation is perhaps the most destabilizing pillar among RRI "keys", while the relationship between participation in R&I processes and ethics is complex. This is due to the diversity of ethical grounds supporting participation, which can vary according to the grounds on which participation is sought (normative/moral justifications) and can either refer to ethics (ethical concerns, ethics broadly understood); to policy (ethics reviews, RRI); or to laws (regulations). The configuration of the involvement, the timescale and the final outcomes of the processes can entail either strong or poor forms of participation. The legitimacy of participation is not self-evident and is often opposed by those who consider that scientific work is already exposed to many constraints both internally and externally (international competitiveness).

The *framework* is aligned with the Sustainable Development Goals (SDGs), following the European Commission's policy approach to innovation, which promotes the "Three Opens": "Open Innovation", "Open Science" and "Open to the World" (Bogers et al., 2018, p. 11). Set as a priority by the European Commission, open innovation responds to the assumption that "we live in a time when those without access to the traditional establishment are often the ones doing the most exciting work"[3]. All forms of participation embedded in R&I processes rely on this premise of the unquestionable added value of any process of involving individual or collective actors not traditionally involved in such processes (referred to as "non-traditional stakeholders"). This assumption is based on the ideal of participative democracy, an overarching normative category encompassing

3 Carlos Moedas' speech "Lund Revisited: Next Steps in Tackling Societal Challenges", Lund, 4 December 2015, published by the European Commission Directorate-General for Research and Innovation (2016).

different approaches, outcomes, activities and timelines[4]: this approach stems from a context of loss of public trust – a democratic deficit particularly salient in the current science–society relationship – where participation springs up as a remedy. Another assumption supporting participatory approaches is that complex public problems can be tackled efficiently by opening channels of participation inviting citizens and stakeholders into public decision-making (Fung, 2008, pp. 681–682), especially considering these voices can be at odds with expert and scientific positions.

Public concern over the impact of emerging technologies is an issue that is embedded in the grounds of participatory processes, with the underlying objective of allowing interested and – directly or indirectly – affected parties to take part in discussions about science, technology and any form of R&I process. *Despite 30 years of participatory technology assessment in Europe, issues with regard to participation such as "why?", "how?" and "in order to reach what quality?" have not yet been solved* (see Pellé and Reber, 2016, Chapter 2), which arguably weakens the legitimacy of participation. The first questions arising with participation in R&I processes are with what purpose, how and, consequently, when does participation occur? In addition, the question of the timeline should be considered, as each stage has a different array of implications, while some participatory activities (e.g. evaluation) can also be considered throughout the whole timeline.

The connection of participation with ethics has several layers of complexity: there are, for instance, several participatory (formalized) procedures, several goals and several connections to responsibility[5]. Differences can be identified in the perspective they offer on ethics; the tools of governance used (ethics committee, participative or deliberative devices, etc.); the understanding of responsibility upon which the actions undertaken rely (often implicitly); and, finally, the characteristics of the inclusion process (e.g. participation or deliberation, the decision-making powers of the stakeholders involved). The legitimacy of participation can be broadly justified through different understandings of responsibility, which, in RRI literature, can be connected to process or outcomes bearing a variety of meanings, both positive and negative ones.

---

4 PRO-Ethics deliverable D1.1, *Report on Mapping of Current Practices of RFOs in Europe*.

5 The following indications for procedures, goals and levels of responsibility mentioned in this paragraph are extensively analyzed in Pellé and Reber (2016).

Overall, there are many different goals for participatory practices, many different approaches to accomplish it and many different ways to connect the outcomes of participation to decision-making. More specifically, there are many different ways to approach participation within the assessment of technologies, laws, research projects and policies, such as: citizen juries, consensus conferences, deliberative conferences, the Delphi and Charette methods, focus groups, planning committees, scenario workshops, "visions of the future" consumer workshops, global cafés, opinion polls (with or without deliberation), questionnaires, citizen advisory committees, vote conferences, interactive TAs, constructivist consumer TAs, ad hoc committees relating to the rules of negotiation, interdisciplinary work groups and political role play, among others. The outcomes of participatory processes in terms of connection to decision-making are also diverse, since reaching an agreement could materialize under consensus, compromise or another form, including consent to disagreement, for instance. In TA, 10 goals for participatory practices can be outlined: assessment of consequences and technical options; extension of perspective for R&D policies; agenda setting; mapping of public scientific controversies; more interactive surveys; coverage of all arguments; reframing of the debate; mediation; policy recommendations (for new technological fields); and comparison of new forms of governance.

Who participates is another key issue in the definition of participation. Usually identified as "citizens" or "stakeholders", these are neither overlapping terms nor do they cover the whole array of potential participants. Commonly identified as an "inclusion" mechanism, participation can embrace various kinds of participants, including stakeholders, citizens, the general public (lay people) and civil society, in addition to the usual participants (scientists and experts). Given the numerous ways of involving citizens or stakeholders in participatory processes for R&I, the questions of who is to be involved, by whom, when and for what also imply a clear identification of the publics.

PRO-Ethics intends to maintain a wide spectrum of participation actors across all dimensions of participatory processes. As a category, "citizens" is broader than "stakeholders" and can ultimately embrace all types of participants in their belonging to a given society. At the same time, "citizens" can also be seen as a category opposed to "stakeholders" as it implies an action toward the common good, ultimately raising the question of the societal purpose or implication of an R&I activity, while other categories can help better frame the quality under which citizens are included in a participatory process (e.g. as end-users or as part of the organized civil society). Who are the publics, either individuals or collectives, that should be associated, in which capacity, how, when and for which outcomes? Defending ethics in the governance of R&I requires adequately addressing the perimeter of ethics and not restraining it to regulatory processes.

Moreover, a lack of definition of the proper process may lead to poor forms of participatory practices. *As there is no single definition of participation, this ethics framework opens up the meanings of participation to embrace the whole array of practices.* Even though there is no single approach toward participative innovation processes, the clear distinction of activities, timelines, expected outcomes and types of participants can provide a common reference. In response to this need, *the common taxonomy and common indicators presented in this ethics framework are part of the guidelines and are meant to be used as a common tool.* In this perspective, reaching a rich form of participatory process requires clarifying several questions, such as which participants should be included or excluded, the nature of the decision-making method (e.g. majority, unanimity, veto right, going as far as possible with argumentation and/or discussion) and the connection to decision-making. Richer forms of participatory processes are to be found in deliberative models, promoting the idea of public participation through iterative processes and fair representation ideals: deliberation is a form of participation where dialog, engagement and justifications are developed to respond to essential democratic requisites, reflecting the theory of deliberative democracy. Deliberation entails the need for adapted approaches to increase the level of deliberation, the proper assessment of the tools and the knowledge shared.

Our research identified several needs facing RFOs. These relate to the definitions of participation and ethics in R&I; the ethical dimensions and ethical issues (as well as their sectoral specificities and how they relate to projects' timelines); the ethical risks and their avoidance; the need for checklists specifying what to consider when involving participants; and the need for general indications regarding ethical challenges, bias and points of attention. In addition to this, specific issues have been identified regarding conflicting interests, methods and the lack of knowledge. At a general level, the distinction between types of engagement actors is essential since citizens and stakeholders, often used interchangeably, are distinct categories, such as citizens as individuals or as part of organized civil society. Following these considerations, participation should be framed through a case-by-case approach, asking the adequate questions and considering the different options in accordance with the context, although some basic requirements, such as informed consent, could be set as a stable basis in a context of responsible innovation, understood as the conduct of responsive processes.

Characteristics of current participatory practices

All European RFOs share a common minimal treatment of ethical aspects. However, not all consider ethics as a primary component of their current participatory processes, owing to a variety of national specificities of a structural and procedural nature and different capacities and competencies. While citizen participation is pursued as a high-priority objective in the discourses and goals of publicly funded R&I, RFOs do not all have the structural means to develop citizen participation activities. As citizen participation mostly occurs before and during R&I, the importance of developing participatory processes after R&I, including social impact assessment, is commonly perceived, although it is challenging to do. The involvement of external stakeholders – in a particular project or policy – can potentially increase the conflicts of interest that RFOs have to cope with, and not all have the competencies required to address this issue. Indeed, the intricacies of stakeholders' interest have to be analyzed, the specific type of actors within the broad category of "stakeholders" needs proper identification and decisions must be made about who is selected to participate and how. Stakeholders could be defined as individuals, groups or organizations who can either affect or be affected by an organization's activities (with a distinction between economic and non-economic stakeholders), while they also represent shared interests. The selection of participants remains a global dilemma that, ultimately, refers to the very framing of the notion of inclusion and democratic representation in our societies.

The added value of participatory processes is another essential feature that is difficult to identify and that meets diverse views (from experts and participants, for instance), the legitimacy being the main guarantee of the sustainability of such processes, although legitimation can be considered at different levels (e.g. law, authority and competences), hence resulting in either broad considerations or specific views. The main characteristics that play an important role in the outcomes of participatory processes are the conditions and goals of genuine participation in regard to decision-making; the means and needs of RFOs, the categorization of modes of participation to outline the perimeter of "participation" in R&I; and the ethical challenges and issues related to participatory approaches that are critical to RFOs (identification and representation of participants, avoidance of biases, use of personal data, etc.). All aforementioned conclusions from the critical review of legal, regulatory and institutional frameworks pertaining to the involvement of citizens and stakeholders were used to develop the specific recommendations and guidelines in this ethics framework.

Part 2 – Tools and guidelines

Considering the inherent complexity resulting from the connection of participation with ethics, how should participation be organized and framed? Rather than providing a list of criteria only, *this ethics framework offers a broad list of questions that should be addressed* and which encompass criteria, classifications and reflections to be undertaken. The purpose of the PRO-Ethics tools and guidelines is to provide a roadmap in the form of questions, following which, the design, the implementation and the evaluation of R&I processes and participatory practices can be tailored in accordance with the specificities of each context. In the criteria, taxonomies and considerations presented below as part of the main questions, there is no hierarchy of participatory processes. The different sections below *address the main ethical considerations and questions helping to determine the participants and the participation modes*: Who? When? How? And what for?

The consideration of these questions is meant to define how ordinary citizens and stakeholders can be invited to participate in R&I processes, both according to an *ethical approach* and with *added value*. In that vein, ethical issues guide these tools, which are comprised of a list of dimensions and questions to address in order to obtain *a roadmap for the diversity of methods and options for participatory approaches*. Considering that there is no universal ideal solution but principles and norms to be contextualized, the purpose of this framework is to provide tools and guidelines to decide whether ethical participation is warranted and what actions and considerations should be undertaken in order to reach it. Since participation is not taken for granted, *this framework suggests taking a step backward and discussing our very assumptions regarding participation*. The most suitable participatory path in each specific case derives from the combination of the context and the specific needs both of the institution undertaking it and of the R&I process that it is applied to.

Below, we offer a set of questions and associated actions to consider when designing, implementing and evaluating a participatory process:

a) How should a participatory process be designed?

b) What type of activity is targeted by the participatory process?

c) What types of participants are targeted?

d) What are the ethics issues and risks?

e) Are selection processes addressing all critical factors?

f) Has the implementation deviated from the design?

g) What are the outcomes and the added value?

Each of the sections includes timeline indications, which are visually highlighted in the box of each specific subset. These indications serve to identify when a specific action is to be undertaken; the timelines may be cumulative in the case of an iterative action (at different stages):

– *before* participation (design phase of the participatory event);

– *during* participation (implementation of the participatory event);

– *after* participation (feedback following the completion of the participatory event).

Also included is a glossary of key terms frequently used in the context of participation in R&I processes to help develop shared understandings and a common language to discuss these topics.

These guidelines are intended to guide the design, implementation and follow-up of participatory processes in accordance with an ethical approach, as a common roadmap encompassing all types of participatory activities and participants. As such, these guidelines may be relevant for researchers, call programmers or scientific/ethics evaluators, for instance. The structure of these guidelines offers common steps and considerations that form a common ground of questions and actions to undertake.

	Guidelines	Timeline of application
Action A	**How should a participatory process be generally structured?**	*[before] [during] [after]*
Subaction A1	***Identify the expected type of contribution and the expected type of input.*** Identify why you are interested in certain types of knowledge and perspectives in the first place. This will allow for the design of more focused discussions and will ensure that the overall intention justifying the need for a participatory process is properly framed and outlined.	*[before]*
Subaction A2	***Allow for flexibility in the planning of the participatory process.*** All stages of R&I processes should be interconnected with the design of the participatory process, anticipating future impacts. Since unexpected nuances and concerns of participants may reveal themselves in the course of the project, flexibility must be considered from the outset, in order to adjust and consider new findings openly. Participatory processes should therefore allocate enough time and resources for their final stages, in order to secure the uptake and incorporation of results and also to secure a meaningful impact. The allocation of resources, time and the selection of participatory methods[6] is intertwined with the extent of flexibility, which should be determined as accurately as possible.	*[before] [during] [after]*
Subaction A3	***Enable a feedback loop and dialog between RFO and R&I projects, as well as between RFO/R&I projects and participants.*** Feedback from concrete experiences in specific fields can provide useful indicators to evaluate where the focus of participatory processes should be placed and how engagement should be furthered in future activities. Transparency toward participants should be established from the outset, by appropriate framing of the overall objectives, the nature and process of the involvement, the expected outcomes, the rights and boundaries of participants and the transparency and accountability commitment on behalf of the organizers of the participatory process. In this perspective, the design of the participatory process should encompass feedback mechanisms.	*[before] [during] [after]*

6 Regarding the diversity of participatory methods to choose from, please refer to action C in this framework.

	Guidelines	Timeline of application
Subaction A4	***Consider the long-term impact of the R&I processes that are considered and the ways to design a participatory process that relate to this concern*** This action should also address concerns of sustainability, of potentially affected people/publics and of long-term unforeseen consequences of social, economic, environmental, cultural, political or scientific nature. Any relevant long-term impact(s) that is/are expected should be listed and further related to the overall design of the participatory process.	*[before]*
Action B	**What type of activity is targeted by the participatory process?**	*[before]*
Subaction B1	***Define the type of activity for which a participatory approach will be undertaken.*** After outlining the above mentioned framing of intent, possibilities, concerns and impacts, these features have to be associated with the identification of the proper participatory activities (see below) in terms of the level and type of interaction that is expected and in view of the overall objectives sought. On a general note, citizens and stakeholders can variously connect to R&I in terms of their contribution and proximity – either internally or externally – to the R&I process. Under this bipartite distinction, three categories of participatory activities can be outlined: – citizen and stakeholder engagement in innovation projects; – citizen and stakeholder engagement within agencies' processes; – citizen and stakeholder engagement in evaluation processes. Within these categories, several types of activities can be outlined, helping to separate the different outcomes of activities involving the participation of citizens: – **general consultation;** – **evaluation** (of projects); – **information/knowledge sharing;** – **monitoring;** – **program design;** – **social impact evaluation;** – **project execution.**	*[before]*

	Guidelines	Timeline of application
Action C	**What types of participants are targeted?**	*[before]*
Subaction C1	***Identify which actors/groups of actors to engage and why, connecting the methods with the context and the type of stakeholders to be involved. Then connect to the previous section (identification of participation activities), so as to cross-check the adequacy.*** The categorization of participants is important both from the perspective of the general distinction of categories and also regarding their proximity with the R&I process[7]. These two specifications allow for better identification of groups that may have been overlooked and different types of participants that may require different treatment. Indeed, the result of the R&I process may be external to the participants' condition (or involvement) or, in other cases (e.g. human participants), the result of the R&I process may directly affect them. The features outlined should be associated with the general identification of the adequate (groups of) participants, according to the nature of the R&I context and the type of participatory activity targeted[8]. This preliminary identification will ensure the methods that are considered are fully compatible with the context and the types of citizens or stakeholders intended to be involved[9].	*[before]*

7 See action B above.

8 See above, action B.

9 In particular, the balance between the roles of researchers and stakeholders should be taken into account, so as to properly address research integrity issues (in the event of conflicting views, for instance).

	Guidelines	Timeline of application
Subaction C2	***Define how a person/group is involved, under what category and the implications (e.g. how to define a person's role as a citizen).*** Following the previous general identification, the actors/groups of actors should be further detailed, reaching the highest level of precision possible. To do so, the following taxonomy can serve to identify which category or categories might be associated and relevant in the given process: – citizens (ordinary people defined as the general public, including citizen scientists and open science movements); – end-users/consumers; – experts; – private body representatives (including institutions, organizations, firms, innovators and entrepreneurs); – public body representatives (including governmental institutions); – organized civil society representatives; – research participants; – scientist and research institution representatives.	*[before]*

	Guidelines	Timeline of application
Action D	**Where are the ethics issues and risks?**	*[before]*
Subaction	***Within each level and timeline phase of the R&I processes, identify the ethical issues and potential R&I risks with an ethical dimension and enhance the structural means within RFOs, so as to be able to tackle them appropriately.*** Once the participatory activity and the participation actors have been defined[10], each of the following potential ethical issues and risks should be considered and further explored if relevant to the given context, in order to determine the feasibility or the participatory process or the need for adaptation. Advice from ethics experts should be sought if any of these issues is deemed relevant. **Consider the following potential issues related to specific R&I processes:** – **In project proposals:** regarding planned products, systems and services and the people involved in projects, issues of human dignity, protection of privacy and data protection, transparency and gender bias should be considered: - potential issues might be: informed consent (integration in/compatibility with the national context); anonymous submission; overrepresentation of certain groups; imbalances in the representation of citizens; insufficient knowledge on the scope of the project; tensions between the scientific quality and the potential benefits of a proposal (during assessment); adequacy and clearness in roles of the panel members; use of panel involvement in the final project selection phase; communication and full information feedback to project leaders. – **In project execution**: issues may related to personal data; discrimination (including gender biases); stigmatization; technology acceptance; dealing with vulnerable groups; distribution of control between humans and technology; privacy; safety; social responsibility of researchers; dealing with social roles in the application context; use of ethically sensitive findings; manipulation and guardianship through technology. – **In evaluation processes:** common ethical risks of legitimacy (due to the lack of formalized ethics review procedures) regarding members' selection, conflicts of interest (such issues also occur regardless of the involvement of citizens); ethical issues in the link to an ethics checklist that is included in the national call text; lack of ethical expertise (at least one evaluator should normally have expertise with respect to ethical issues); ethical issues in promoting the call by specifically addressing ethical issues in events. – **In agency processes:** ethical issues might appear in the call text (program design); in the planning of one-way engagement activities (information/knowledge sharing/communication) which, in their lowest forms, could be unduly considered as participatory; in the impact of participation, depending on the proper interpretation and understanding of the context in which it is conducted; in the degree of development of the ethical reflection and the ethical competencies.	*[before]*

10 See actions B and C, respectively.

	Guidelines	Timeline of application
Subaction	**Consider the following issues that may occur at the general level:** – ***Conflicting interests:*** - ensure there are no conflicts of interest (conflicts with existing funding structures and processes or internal conflicts related for instance to doubts regarding the planned participatory process); - ensure that citizens and/or stakeholders' participation also includes representation of diverse voices and voices at odds; - identify potential issues of legal capacity (impartiality, partiality, external conflicts). – ***Methods:*** - take into consideration that when participation is made a mandatory requirement for funded R&I projects, this raises the hurdle for new institutions to access funding; - identify the level of adequacy of the selected participatory process, in regard to: (1) whether participation is warranted in the given project and (2) whether the involvement of participants would benefit from adequate support; - in the case of top-down funding calls, reflect on the possibility of having a participatory topic identification; - ensure the selection of participatory approaches clearly answers the following questions: who? when? how? (see previous sections). – ***Lack of knowledge/awareness:*** - since the option of participatory processes is sometimes not taken into account because of a lack of information (either at the institutional or at citizen and stakeholder level), a dissemination strategy and the identification of adequate resources to inform potential participants should be considered; - identify the level of knowledge needed (scientific/technical background) for the execution and future evaluation of the participatory activity, in each specific context/sector considered, and ensure that potential ethical problems that have been spotted find the required expertise. – ***Scientific integrity:*** - identify if (and how) the engagement might affect the researchers; - ensure the methodology of participation and the reporting on the outcomes of participation are properly adapted to the values of research integrity; - ensure the ability – at RFO level – to properly relate the participatory process to the scientific community and their values.	***[before]***

	Guidelines	Timeline of application
Action E	**Do the selection processes address all critical factors?**	*[before]*
Subaction	***During the design of a participatory process, take into consideration the following sensitive factors and adapt them according to national contexts.*** Further to the ethical risks and issues[11], this action ensures that the criteria to implement fair and gender-sensitive ethical procedures within new interaction modes are considered so as to: – provide more opportunities and access to innovation for those who are excluded, which involves reflecting on the proper balance between pluralism and representation (proportionality); this may induce potential compromises to reach inclusion, as well as potential overrepresentation of minorities; – properly determine when a category of people is included and who exactly is excluded; – clarify the process of selection, the socio-demographic criteria used, the justification of the choice and the categories of participants (referring to the above mentioned list). **Selection of citizens and stakeholders:** – reflect on the initial identification of participants[12] and the way they effectively relate to this assigned category, in case a change of category is needed; – for each category of participants (e.g. citizens or scientists), identify which knowledge background is expected from them and with which they will be asked to act;	*[before]*

11 See above, action D.

12 See above, action C.

	Guidelines	Timeline of application
Subaction	– ensure there have not been biases in the selection of participants and that there is broad representation; – consider potential issues of asymmetric access to information on behalf of citizens and make sure such issues can be tackled; – put in balance the potential benefits and challenges of different selection processes considered (diversity of participants, size of the group) and relate them to the project's objectives or subject; – ensure the collection and processing of personal data is addressed according to the applicable national and European legislation; – ensure vulnerable groups are meaningfully involved/engaged; – identify the right segment of stakeholders that should participate and the purpose of their participation and avoid the involvement of participants whose contribution is not properly identified; – in the case of topic selection, consider the required level of participants'[13] expertise and make any necessary adjustments prior to the final design of the participatory process; – identify all potential stakeholders in regard to the context, sector, project/program needs, expected outcomes and representativity issues to include voices at odds; – anticipate the potential clashes between certain types of participants, or between participants and the participatory methods to be used, and identify the methods to properly tackle and overcome them; – consider whether a digital or analog manner for involvement is more appropriate.	***[before]***

13 This relates to the expertise that is expected on behalf of participation actors (citizens or stakeholders – see above, action C, for the related taxonomy) in view of their contribution to the participatory activity.

	Guidelines	Timeline of application
Action F	**Has the implementation deviated from the design?**	*[after]*
Subaction	***After an R&I process that has included citizen or stakeholder participation, consider the following questions and criteria and, if they are not met, modify future RFO participatory initiatives.*** – Participants of engagement activities should be first broadly identified, then categorized according to the taxonomies (action C, above) and the category of activity and the level of action (action B, above), aiming to reach the highest level of detail both in regard to participants and activities and to identify any potential deviation during implementation from the initial design[14]. – Depending on the outcomes of the participatory process, while some inputs may be chosen to be disregarded, the overall accountability process should consider all voices and keep track of those excluded from the final outputs. – Verify whether representation and inclusion of participants have been addressed throughout the whole timeline of the R&I process in order to identify potential deviations in the adequacy of their selection or in their treatment. – Ensure that the actual input from citizens and stakeholders is taken into consideration in the final decisions following a participatory process. – The societal justification of participation should be taken into account as it ensures that the quality and functionality of procedures connects with societal needs.	*[after]*

14 This distinction has an impact on the overall participatory approach: see indications in actions B and C.

	Guidelines	Timeline of application
Action G	**What are the outcomes and the added value?**	***[before] [during] [after]***
Subaction G1	***Before and during implementation of the participatory process, consider the additional resources needed for participatory approaches, which need to be balanced against the expected benefits.*** Participatory processes should be accompanied by sufficient resources to ensure beneficial outcomes. Moreover, the added value of a participatory process is partly determined by the balance between the expected benefits and the costs and resources effectively dedicated to the participatory process. This balance should be determined as precisely as possible from the design stages of the participatory process. This should be addressed both before and during implementation of the participatory process. The added value should also relate to the specificities of the contribution of the selected participants (e.g. citizens) and the reasons why the chosen category has provided a contribution that could not have been provided by other types of participants.	***[before] [during]***
Subaction G2	***Before and during implementation, and on the basis of the previous action, establish a timeline for the monitoring of participation outcomes.*** This reflection should mention: (1) which outcomes are expected to be kept and integrated into R&I/RFO processes; and (2) which outcomes are not expected to be kept (and the reason why). During implementation, these expectations should be reconsidered and adapted if needed, following reflection on the deviation from pre-set indicators in the tangible results of the participatory process[15].	***[before] [during]***
Subaction G3	***After the completion of the integration of participation outcomes, launch a transparency process allowing participants to interact.*** According to the size of the participatory activity and the organizational capabilities, participants may be contacted to receive information on the treatment of their participation (e.g. data protection), the specific features that have been kept or rejected and the overall objectives met through the outcomes of the participatory process and to allow them to interact. This feedback from participants should be collected as a main source of assessment on the process and should be considered in future stages of participatory processes, indicating potential needs for adaptation. An ideal implementation of this action would be to officially acknowledge the right of citizens (or stakeholders) to assess the outcomes of their proposals/inputs and to request accountability provisions.	***[after]***

15 This action is complementary with subactions A2 and A3 (see above, action A).

	Guidelines	Timeline of application
Subaction G4	***Establish a form of compensation/acknowledgment for the participants' contribution.*** The result of this action should ensure participants feel their contribution has been considered and valued. This could consist of a financial retribution or an official acknowledgment, depending on the possibilities in each specific context.	*[after]*
Subaction G5	***Establish a list of benefits from the participatory process that could not have been achieved without the involvement of the specific participant(s) or group(s) of participants.*** This action serves to verify the assumption that participation is justified in all cases, regardless of the reasons and approaches.	*[after]*
Subaction G6	***Both in view of future reference and for accountability matters, all reflections addressing the actions mentioned in these guidelines should be kept in a synthetic report, to be archived by the RFO and reviewed by its board members at a later stage.*** Keeping track of all aforementioned actions related to the participatory processes will enable appropriate treatment of transparency and accountability issues that may arise at a later stage, and should also serve for any future retrospective analysis of the process(es) undertaken.	*[after]*

	Guidelines	Timeline of application
Action H	***Using common definitions***	***[before] [during] [after]***
	As well as general terms of interest, the following glossary encompasses terms that are frequently used to describe: – how a person/group is involved, under what category and with what implications; – the type and level of activity of participatory processes; – the distinct levels of interactivity/degrees of involvement of society in research. The categories and definitions outlined below reflect the work undertaken in the PRO-Ethics project and serve as common references. *Some of these categories are part of the "classifications" in the above tools and guidelines* section of the present ethics framework; others are additional terms and definitions of typical concepts and notions that are commonly used to describe ethics and participation within R&I processes. **Citizens** This category includes the general public, lay people and citizens as persons (or collectives) with civic expectations[16]. Moreover, since end-users can be categorized as citizens as well, this distinction serves to underline the general dimension of involvement, referring to the broader sense of "public participation". This category may also relate to a broad participatory approach. **Consultation** Processes of engagement with any group of citizens or stakeholders in order to obtain feedback on policies and programs.	

16 In this category, the main boundaries for the "citizens" category are anchored in the citizens/stakeholders distinction.

	Guidelines	Timeline of application
	End-users/consumers End-users/consumers are defined as a specific category distinct from citizens (sometimes with overlaps) referring to beneficiaries of the end product (including solutions and services) of the R&I process. This category specifically refers to the impact and the uses of the products of R&I activities in society. **Engagement** Information and knowledge about R&I activities can be provided and disseminated (e.g. dissemination of research to public), thus raising awareness about R&I activities (e.g. through media, science festivals and open days at universities and research centers)[17]. **Ethics and integrity** Integrity refers to research findings and the process in which they are produced (i.e. data, methods, interpretation and presentation/reporting) and whether such processes and findings meet established and appropriate scientific, legal and professional standards. In comparison, "research ethics" pertains to the moral issues that occur in the research design and its implementation, for instance in relation to the protection of humans, animals, the environment and data, as well as the proper protection of other objects[18]. **Evaluation (of projects and programs)** This category encompasses several types of evaluation: evaluation of project proposals (i.e. the ethical and scientific evaluation) as part of the selection process intervening in funding schemes; interim and *ex post* evaluation for projects and programs that received funding; and program evaluation.	

17 Source: https://www.invo.org.uk.

18 Source: Ravn and Braun, 2019.

	Guidelines	Timeline of application
	Experts This category serves to identify individuals enrolled as (internal or external) experts in RFO activities, but at the individual level and not as part of any of the other categories. Therefore, lay experts for instance can be included in this category, if their involvement mainly values their expertise (as citizens) and if this category prevails over others (e.g. "citizens"). Experts can be individuals with any sectoral expertise (e.g. with a background in medicine, psychology, sociology or philosophy, among others disciplines). **Funding budget** Most RFOs distinguish between "commitments", for example allowances for new projects, and "liquidations" , for example payments for current projects. In the context of PRO-Ethics, we always refer to the commitment budget. **Information/knowledge sharing** Dissemination of information about R&I contents, outputs and scientific knowledge. Also one-way like the communication of funding activities, knowledge sharing relates to the dissemination of scientific content only. This mostly relates to one-way engagement toward any type of public and the sharing of information related to communication of funding activities (funding portfolio) and/or funded activities (the funded research, development and innovation and its results). **Involvement** Citizens and stakeholders can be actively involved in R&I activities: for instance, as patients, as grant holders and co-applicants, in identifying research opportunities, in agenda setting, as members of project advisory and steering groups, in the co-developing of patient information or materials or in undertaking interviews with participants and carrying out research[19].	

19 Source: https://www.invo.org.uk/.

	Guidelines	Timeline of application
	Monitoring The systematic follow-up of funded projects in the context of RFO funding schemes is usually an activity carried out internally (but which can be supported by external experts: e.g. when doing interim or final reviews), although *ex post* monitoring of results can involve other actors, in addition to the involvement (feedback) of programs' beneficiaries. **Organized civil society** Civil society organizations have different knowledge and leverage compared to individual citizens. They may defend interests, often professional interests (trade unions), causes (e.g. animals and environmental issues) or rights (e.g. minorities and women). **Participants** Participants are defined as persons who take part in participative processes. These persons might be: citizens without a specific interest in the case (referred to as "citizen participation"); (end-)users with a specific interest in the results (also part of "citizen participation"); or stakeholders including non-traditional stakeholders such as NGOs with specific knowledge and/or specific interests (not part of "citizen participation") and with any level of engagement and interaction. Furthermore, participants can be either individuals or representatives of institutions or groups (organized civil society) and may include vulnerable groups such as patients, children or older adults. In each case, the kinds of participating persons should be carefully described. **Participation** The process of citizens and patients taking part in research and innovation studies (e.g. as patients being recruited in clinical trials; by completing questionnaires; as participants in interviews and focus groups)[20]. **Private body representatives (innovators/entrepreneurs)** Representatives of SMEs or large companies (including those conducting research), considered as beneficiaries (actors who implement the funded proposals) or non-beneficiaries of funding schemes and involved in some way in RFO activities as participants, participating on behalf of their company. This category also comprises social entrepreneurs.	

20 Source*:* https://www.invo.org.uk/.

	Guidelines	Timeline of application
	Program design In the context of research and innovation, program design refers to the identification of program objectives and of R&I priorities, resulting in the definition of funding opportunities. **Project lead** The project lead may be a principal investigator (PI) or a coordinating person/institution/enterprise responsible for the whole R&I project. **Public body representatives (governmental institutions)** Also understood as governmental institutions, this category encompasses national public bodies (e.g. government departments or ministries, excluding academic research institutions) that might be involved in the management of RFO activities and projects or that might operate as funding agencies. Public bodies relate to ethics through ethical compliance mainly, ensuring ethics requirements are met in grant agreements signed with the beneficiaries at a national level. **R&I (or RDI) agents** Research and innovation (or research, development and innovation) agents include researchers in higher education institutions (HEIs) and research performing organizations, as well as researchers, developers and innovators in enterprises, NGOs and public bodies. As in the case of participants, RDI agents might be individuals (e.g. individual researchers) or institutions/institutional delegates (e.g. universities or firms) carrying out the research as beneficiaries. **Research participants** This category refers to human participants who are part of the subject of research (for instance, in medical sciences, public and patients' involvement).	

	Guidelines	Timeline of application
	RFO activities In the context of PRO-Ethics, RFO activities (processes) refer to program/funding scheme development and implementation (launch of the call, reception of proposals, proposal selection), as well as strategic planning (including grant/program management). The processes might also refer to, for example, organizational concepts and personnel development but proposal selection is a core process of RFOs and is thus treated as a separate category. **Scientists/research institution representatives** Individual scientists (any scientific or technical field) or representatives of research institutions involved in RFO processes either for their individual expertise or for their affiliation to a research institution, but speaking for themselves only. If they are invited on behalf of a private company conducting research, these participants should be identified as part of the private body representatives category. This category also includes researchers as physical persons working for a legal entity (SMEs, universities, research institutes) conducting research and to whom public or private funding has been granted to do so. Participants under this category can be considered experts. **Social impact evaluation** RFO activities that aim at measuring the societal impact of R&I processes, including the development of indicators and monitoring allowing such measurement.	

References

AccountAbility (2015). AA1000 Stakeholder Engagement Standard (AA1000SES). AccountAbility, London [Online]. Available at: https://www.accountability.org/standards/aa1000-stakeholder-engagement-standard/ [Accessed 31 January 2022].

All European Academies (2017). The European Code of Conduct for Research Integrity, revised edition. ALLEA, Berlin.

All European Academies and European Science Foundation (2011). The European Code of Conduct for Research Integrity. Ireg, Strasbourg.

Arnstein, S.R. (1969). A ladder of citizen participation. *Journal of the American Planning Association*, 35(4), 216–224.

Asveld, L. and van Dam-Mieras, R. (2017). Introduction: Responsible research and innovation for sustainability. In *Responsible Innovation 3: A European Agenda?*, Asveld, L., van Dam-Mieras, R., Swierstra, T., Lavrijssen, S., Linse, K., Van den Hoven, J. (eds). Springer International Publishing, Cham.

Bergold, T. and Thomas, S. (2012). Participatory research methods: A methodological approach in motion. *Forum Qualitative Sozialforschung/Forum: Qualitative Social Research*, 13(1), Art. 30.

Bessant, J. (2013). Innovation in the twenty-first century. In *Responsible Innovation*, Owen, R., Heintz, M., Bessant, J.R. (eds). John Wiley & Sons, Chichester.

Blok, V., Hoffmans, L., Wubben, E.F.M. (2015). Stakeholder engagement for responsible innovation in the private sector: Critical issues and management practices. *Journal on Chain and Network Science*, 15(2), 147–164.

Bogers, M., Chesbrough, H., Moedas, C. (2018). Open innovation: Research, practices, and policies. *California Management Review*, 60(2), 5–16.

Brom, F.W.A., Chaturvedi, S., Ladikas, M., Zhang, W. (2015). Institutionalizing ethical debates in science, technology, and innovation policy: A comparison of Europe, India and China. In *Science and Technology Governance and Ethics*, Ladikas, M., Chaturvedi, S., Zhao, Y., Stemerding, D. (eds). Springer, Cham.

Chambers, S. (2003). Deliberative democratic theory. *Annual Review of Political Science*, 6, 309.

Chesbrough, H. (2006). Open innovation: A new paradigm for understanding industrial innovation. In *Open Innovation: Researching a New Paradigm*, Chesbrough, H., Vanhaverbeke, W., West J. (eds). Oxford University Press, Oxford.

Chilvers, J. (2010). Sustainable participation? Mapping out and reflecting on the field of public dialogue on science and technology. Summary report, Sciencewise-ERC and the University of East Anglia, London.

Council of Europe (2012). Guide for Research Ethics Committee Members. Steering Committee on Bioethics. Council of Europe, Strasbourg [Online]. Available at: https://rm.coe.int/CoERMPublicCommonSearchServices/DisplayDCTMContent?documentId=0900001680307e6c [Accessed 31 January 2022].

Crozier, M. and Friedberg, E. (1993). *Die Zwänge kollektiven Handelns – Über Macht und Organisation*. Hain, Frankfurt.

Cuppen, E., Klievink, B., Doorn, N. (2019). Governing crowdbased innovations: An interdisciplinary research agenda. *Journal of Responsible Innovation*, 6(2), 232–239.

Dahlman, C. and Kuznetsov, Y. (2014). Innovation for the "base of the pyramid": Developing a framework for policy experimentation. In *Making Innovation Policy Work: Learning from Experimentation*. The World Bank, Washington, D.C., and OECD Publishing, Paris.

Davidson, S. (1998). Spinning the wheel of empowerment. *Planning*, 1262, 14–15.

Demeritt, D., Dobson, A., Murray Li, T., Leach, M., Scoones, I., Stirling, A. (2011). Pathways to sustainability: Perspectives and provocations. *Environment and Planning*, 43, 1226–1237.

Doorn, N. and Nihlén Fahlquist, J.A. (2010). Responsibility in engineering: Towards a new role for engineering ethicists. *Bulletin of Science, Technology and Society*, 30, 222–230.

Driessen, A. (2009). Ethical aspects of research in ultrafast communication. In *Evaluating New Technologies*, Sollie, P. and Düwell, M. (eds). Springer, Dordrecht.

Dutz, M., Kuznetsov, Y., Lasagabaster, E., Pilat, D. (eds) (2014). *Making Innovation Policy Work: Learning from Experimentation*. The World Bank, Washington, D.C., and OECD Publishing, Paris.

Edwards, M.A. and Siddhartha, R. (2017). Academic research in the 21st century: Maintaining scientific integrity in a climate of perverse incentives and hypercompetition. *Environmental Engineering Science*, 34(1), 51–61.

Esterling, K.M., Fung, A., Lee, T. (2015). How much disagreement is good for democratic deliberation? *Political Communication*, 32(4), 529–551.

European Commission (2013). The EU Framework Programme for Research and Innovation.

European Commission (2018). Ethics and data protection [Online]. Available at: https://ec.europa.eu/info/sites/info/files/5._h2020_ethics_and_data_protection_0.pdf [Accessed 31 January 2022].

European Commission Directorate-General for Research and Innovation (2010). *European Textbook on Ethics in Research*. Publications Office of the European Union, Luxembourg.

European Commission Directorate-General for Research and Innovation (2013). *Ethics for Researchers: Facilitating Research Excellence in FP7*. Publications Office of the European Union, Luxembourg [Online]. Available at: https://ec.europa.eu/research/participants/data/ref/fp7/89888/ethics-for-researchers_en.pdf [Accessed 31 January 2022].

European Commission Directorate-General for Research and Innovation (2016). *Open Innovation, Open Science, Open to the World*. Publications Office of the European Union, Luxembourg.

European Commission Directorate-General for Research and Innovation (2019). Horizon 2020 Programme: Guidance – How to complete your ethics self-assessment [Online]. Available at: https://ec.europa.eu/research/participants/data/ref/h2020/grants_manual/hi/ethics/h2020_hi_ethics-self-assess_en.pdf [Accessed 31 January 2022].

European Group on Ethics in Science and New Technologies (2015). *The Ethical Implications of New Health Technologies and Citizen Participation*. Publications Office of the European Union, Luxembourg [Online]. Available at: https://op.europa.eu/en/publication-detail/-/publication/e86c21fa-ef2f-11e5-8529-01aa75ed71a1 [Accessed 31 January 2022].

European Parliament and European Council (2014). Regulation (EU), No. 536/2014 on clinical trials on medicinal products for human use, and repealing Directive 2001/20/EC. *Official Journal of the European Union*, L158, 1–76 [Online]. Available at: https://eur-lex.europa.eu/legal-content/EN/TXT/?uri=celex%3A32014R0536 [Accessed 31 January 2022].

Fagotto, E. and Fung, A. (2014). Embedding public deliberation in community governance. In *Varieties of Civic Innovation: Deliberative, Collaborative, Network, and Narrative Approaches*, Girouard, J. and Sirianni, C. (eds). Vanderbilt University Press, Nashville.

Fecher, B. and Friesike, S. (2013). Open science: One term, five schools of thought. In *Opening Science*, Friesike, S. and Bartling, S. (eds). Springer, Cham.

Felin, T. and Zenger, T.R. (2014). Closed or open innovation? Problem solving and the governance choice. *Research Policy*, 43(5), 914–925.

Fiorino, D.J. (1990). Citizen participation and environmental risk: A survey of institutional mechanisms. *Science, Technology, and Human Values*, 15(2), 226–243.

Fisher, E., Mahajan, R.L., Mitcham, C. (2006). Midstream modulation of technology: Governance from within. *Bulletin of Science, Technology and Society*, 26(6), 485–496.

Fishkin, J. (2009). *When the People Speak: Deliberative Democracy and Public Consultation*. Oxford University Press, Oxford.

Francis, D. and Bessant, J. (2005). Targeting innovation and implications for capability development. *Technovation*, 25(3), 171–183.

Fung, A. (2003). Associations and democracy: Between theories, hopes, and realities. *Annual Review of Sociology*, 291(1), 515–539.

Fung, A. (2008). Democratizing the policy process. In *The Oxford Handbook of Public Policy*, Goodin, R.E. (ed.). Oxford University Press, Oxford.

Gianni, R. (2016). *Responsibility and Freedom. The Ethical Realm of RRI.* ISTE Ltd, London, and John Wiley & Sons, New York.

Granqvist, K. (2016). Policy Brief: Funding Frugal Innovation. Lessons on Design and Implementation of Public Funding Schemes Stimulating Frugal Innovation. Centre for Social Innovation (ZSI), Vienna.

Griessler, E., Lang, A., Wuketich, M. (2015). Analytical report on the dimension of research and innovation ethics. MoRRI, Deliverable D2.4.2. [Online] Available at: https://morri.netlify.app [Accessed 12 April 2022].

Grinbaum, A. and Groves, C. (2013). What is "responsible" about responsible innovation? Understanding the ethical issues. In *Responsible Innovation*, Owen, R., Heintz, M., Bessant, J.R. (eds). John Wiley & Sons, Chichester.

Guchet, X. (2021). *Care in Technology*. ISTE Ltd, London, and John Wiley & Sons, New York.

Guston, D.H. and Sarewitz, D. (2002). Real-time technology assessment. *Technology in Society*, 24(1), 93–109.

Gutman, A. and Thompson, D. (2004). *Why Deliberative Democracy?* Princeton University Press, Princeton.

Hossain, M. (2016). Grassroots innovation: A systematic review of two decades of research. *Journal of Cleaner Production*, 137, 973–981.

Houllier, F. and Merilhou-Goudard, J.-B. (2016). Les sciences participatives en France. État des lieux, bonnes pratiques et recommandations. Report [Online]. Available at: http://www.sciences-participatives.com/Rapport [Accessed 31 January 2022].

Howaldt, J. and Schwarz, M. (2010). Social innovation: Concepts, research fields and international trends. *IMO International Monitoring, 5*.

Howaldt, J., Domanski, D., Kaletka, C. (2016). Social innovation: Towards a new innovation paradigm. *Revista De Administração Mackenzie*, 17(6), 20–44.

Humanitarian UAV Network (2015). Humanitarian UAV code of conduct [Online]. Available at: https://uavcode.org [Accessed 31 January 2022].

Institute of Electrical and Electronics Engineers (2006). IEEE policies: Code of ethics [Online]. Available at: https://www.ieee.org/about/corporate/governance/p7-8.html [Accessed 31 January 2022].

International Association for Public Participation (2018). IAP2's spectrum of public participation [Online]. Available at: https://cdn.ymaws.com/www.iap2.org/resource/resmgr/pillars/Spectrum_8.5x11_Print.pdf [Accessed 31 January 2022].

Irwin, A. (2006). The politics of talk: Coming to terms with the "new" scientific governance. *Social Studies of Science*, 36(2), 299–320.

Jasanoff, S. (2005). *Designs on Nature: Science and Democracy in Europe and the United States*. Princeton University Press, Princeton.

Kaplinsky, R. (2014). "Bottom of the pyramid" innovation and pro-poor growth. In *Making Innovation Policy Work: Learning from Experimentation*, Dutz, M., Kuznetsov, Y., Lasagabaster, E., Pilat, D. (eds). The World Bank, Washington, D.C., and OECD Publishing, Paris.

Karatzas, I. (2015). The Ethics Appraisal Scheme in Horizon 2020 [Online]. Available at: https://bestprac.eu/fileadmin/mediapool-bestprac/documents/Outputs/Learning_materials/Ethics_H2020_Appraisal_2017_Karatzas.pdf [Accessed 31 January 2022].

Khan, S.S., Timotijevic, L., Newton, R., Coutinho, D., Llerena, J.L., Ortega, S., Benighaus, L., Hofmaier, C., Xhaferri, Z., de Boer, A., Urban, C., Strähle, M., Da Pos, L., Neresini, F., Raats, M.M., Hadwiger, K. (2016). The framing of innovation among European research funding actors: Assessing the potential for "responsible research and innovation" in the food and health domain. *Food Policy*, 62, 78–87.

Klüver, L., Bellucci, L.S., Bütschi, D., van Eijndhoven, J., van Est, R., Gloede, F., Hennen, L., Joss, S., Nentwich, M., Peissl, W., Torgersen, H. (2000). *EUROpTA: European Participatory Technology Assessment – Participatory Methods in Technology Assessment and Technology Decision-Making*. Danish Board of Technology, Copenhagen.

Kretser, A., Murphy, D., Bertuzzi, S., Abraham, T., Allison, D.B., Boor, K.J., Dwyer, J., Grantham, A., Harris, L.J., Hollander, R., Jacobs-Young, C., Rovito, S., Vafiadis, D., Woteki, C., Wyndham, J., Yada, R. (2019). Scientific integrity principles and best practices: Recommendations from a scientific integrity consortium. *Science and Engineering Ethics*, 25(2), 327–355.

Kroll, H., Gabriel, M., Braun, A., Muller, E., Neuhäusler, P., Schnabl, E., Zenker, A. (eds) (2016). A conceptual analysis of foundations, trends and relevant potentials in the field of frugal innovation (for Europe). Interim Report for the Project "Study on Frugal Innovation and Reengineering of Traditional Techniques", Fraunhofer ISI and Nesta, Publications Office of the European Union, Luxembourg.

Kurt, A., Duquenoy, P., Lavelle, S. (2010). Ethical governance models, paradigm recognition and interpretation. EGAIS Deliverable D3.1.

Laursen, K. and Salter, A.J. (2014). The paradox of openness: Appropriability, external search and collaboration. *Research Policy*, 43(5), 867–878.

Lenoir, C.L. (2019). *Ethically Structured Processes*. ISTE Ltd, London, and John Wiley & Sons, New York.

Loisy, C. and Coulet, J.-C. (eds) (2018). *Competence and Program-based Approach in Training*. ISTE Ltd, London, and John Wiley & Sons, New York.

Maesschalck, M. (2017). *Reflexive Governance for Research and Innovative Knowledge*. ISTE Ltd, London, and John Wiley & Sons, New York.

Manders-Huits, N. and Van den Hoven, M.J. (2009). The moral relevance of technological artifacts. In *Evaluating New Technologies*, Sollie, P. and Düwell, M. (eds). Springer, Dordrecht.

Mansbridge, J. (2014). A systemic approach to civic action. In *Varieties of Civic Innovation: Deliberative, Collaborative, Network, and Narrative Approaches*, Girouard, J. and Sirianni, C. (eds). Vanderbilt University Press, Nashville.

Mejlgaard, N. and Ravn, T. (2015). Monitoring the evolution and benefits of responsible research and innovation. MoRRI Deliverable D2.1 [Online] Available at: https://morri.netlify.app [Accessed 12 April 2022].

Monaghan, A. (2009). Conceptual niche management of grassroots innovation for sustainability: The case of body disposal practices in the UK. *Technological Forecasting and Social Change*, 76, 1026–1043.

Mouffe, C. (1999). Deliberative democracy or agonistic pluralism? *Social Research*, 66(3), 745–758.

Moulaert, F., Mehmood, A., MacCallum, D., Leubolt, B. (eds) (2017). *Social Innovation as a Trigger for Transformations: The Role of Research*. Publications Office of the European Union, Luxembourg.

Mulgan, G., Ali, R., Halkett, R., Sanders, B. (2007a). *In and Out of Sync. The Challenge of Growing Social Innovations*. Nesta, London.

Mulgan, G., Tucker, S., Ali, R., Sanders, B. (2007b). *Social Innovation – What It Is, Why It Matters and How It Can be Accelerated*. The Young Foundation, London.

Murray, R., Caulier-Grice, J., Mulgan, G. (2010). *The Open Book of Social Innovation*. The Young Foundation and Nesta, London.

Nascimento, S., Iglesias, J.M.R., Owen, R., Schade, S., Shanley, L. (2018). Citizen science for policy formulation and implementation. In *Citizen Science: Innovation in Open Science, Society and Policy*, Hecker, S., Haklay, M., Bowser, A., Makuch, Z., Vogel, J., Bonn, A. (eds). UCL Press, London.

National Commission for the Protection of Human Subjects of Biomedical and Behavioral Research (1979). The Belmont Report: Ethical principles and guidelines for the protection of human subjects of research [Online]. Available at: https://www.hhs.gov/ohrp/regulations-and-policy/belmont-report/read-the-belmont-report/index.html [Accessed 31 January 2022].

OECD (2015). *Innovation Policies for Inclusive Growth*. OECD Publishing, Paris.

OECD and Statistical Office of the European Communities (2018). *Oslo Manual: Guidelines for Collecting, Reporting and Using Data on Innovation*, 4th edition. OECD, Paris.

Official Journal of the European Union (2010a). Charter of Fundamental Rights of the European Union. No. C 83/393.

Official Journal of the European Union (2010b). Consolidated Version of the Treaty on the Functioning of the European Union. No. 83/55.

Oldsman, E. and Nexus Associates (2014). Making evaluations count: Toward more informed policy. In *Making Innovation Policy Work: Learning from Experimentation*, Dutz, M., Kuznetsov, Y., Lasagabaster, E., Pilat, D. (eds). The World Bank, Washington, D.C., and OECD Publishing, Paris.

Owen, R. and Pansera, M. (2019). Responsible innovation and responsible research and innovation. In *Handbook on Science and Public Policy*, Simon, D., Kuhlmann, S., Stamm, J., Canzler, W. (eds). Edward Elgar Publishing, Cheltenham.

Owen, R., Macnaghten, P., Stilgoe, J. (2012). Responsible research and innovation: From science in society to science for society, with society. *Science and Public Policy*, 39(6), 751–760.

Owen, R., Heintz, M., Bessant, J.R. (eds) (2013). *Responsible Innovation: Managing the Responsible Emergence of Science and Innovation in Society*. John Wiley & Sons, Chichester.

Patel, M., Sotsky, J., Gourley, S., Houghton, D. (2013). *The Emergence of Civic Tech: Investments in a Growing Field*. The Knight Foundation, New York.

Pellé, S. (2017). *Business, Innovation and Responsibility*. ISTE Ltd, London, and John Wiley & Sons, New York.

Pellé, S. and Reber, B. (2016). *From Ethical Review to Responsible Research and Innovation*. ISTE Ltd, London, and John Wiley & Sons, New York.

Pervez, T. and Maritz, A. (2013) Innovation and social entrepreneurship at the bottom of the pyramid – A conceptual framework. *South African Journal of Economic and Management Sciences*, 16(5), 54–66.

Peter, V. and Maier, F. (eds) (2018). *Monitoring the Evolution and Benefits of Responsible Research and Innovation in Europe – Summarising Insights from the MoRRI Project*. Publications Office of the European Union, Luxembourg.

Peter, V., Maier, F., Spaini, C., Woolley, R., Meijer, I., Costa, R., Bloch, C., Mejlgaard, N. (2018). *Monitoring the Evolution and Benefits of Responsible Research and Innovation in Europe. The Evolution of Responsible Research and Innovation – The Indicators Report*. Publications Office of the European Union, Luxembourg.

Powell, M. and Colin, M. (2009). Participatory paradoxes. Facilitating citizen engagement in science and technology from the top-down? *Bulletin of Science, Technology and Society*, 29(4), 325–342.

Rask, M., Mačiukaitė-Žvinienė, S., Tauginiene, L., Dikčius, V., Matschoss, K., Aarrevaara, T., d'Andrea, L. (2018). *Public Participation, Science and Society: Tools for Dynamic and Responsible Governance of Research and Innovation*. Routledge, New York.

Ravn, T. and Braun, R. (eds) (2019). RE/RI expert set of indicators for e-database. ENERI Deliverable D6.2 [Online] Available at: https://eneri.eu/wp-content/uploads/2019/10/D6.2-Evaluation-Report-on-Pilot-Set-of-Indicators.pdf [Accessed 12 April 2022].

Reber, B. (2005). Technologies et débat démocratique en Europe. *Revue française de science politique*, 5, 811–833.

Reber, B. (2006). Technology assessment as policy analysis: From expert advice to participatory approaches. In *Handbook of Public Policy Analysis. Theory, Politics, and Methods*, Fischer, F., Miller, G., Sidney, M. (eds). Rutgers University and CRC Press, New York.

Reber, B. (2016). *Precautionary Principle, Pluralism and Deliberation: Science and Ethics*. ISTE Ltd, London, and John Wiley & Sons, New York.

Reber, B. (2018). RRI as the inheritor of deliberative democracy and precautionary principle. *Journal of Responsible Innovation: Responsible Innovation and Brain*, 5(1), 38–64.

Reber, B. (2019). Taking moral responsibility seriously to foster responsible research and innovation. In *Responsible Research and Innovation: From Concepts to Practices*, Gianni, R., Pearson, J., Reber, B. (eds). Routledge, Oxford.

Reber, B. (2022). *Responsible Deliberation Between Conversation and Consideration*. ISTE Ltd, London, and John Wiley & Sons, New York [to be published].

Reed, M. (2008). Stakeholder participation for environmental management: A literature review. *Biological Conservation*, 141(10), 2417–2431.

Richardson, A. (1983). *Participation*. Routledge, London.

Robinson, L.D., Cawthray, J.L., West, S.E., Bonn, A., Ansine, J. (2018). Ten principles of citizen science. In *Innovation in Open Science, Society and Policy*, Hecker, S., Haklay, M., Bowser, A., Makuch, Z., Vogel, J., Bonn, A. (eds). UCL Press, London.

Rodrigues, R., Broadhead, S., Brey, P., Warso, Z., Hanson, T., Tambornino, L., Lanzerath, D. (2018). SIENNA: The consortium's methodological handbook. SIENNA Deliverable D1.1. [Online] Available at: https://www.sienna-project.eu/publications/deliverable-reports/ [Accessed 12 April 2022].

Rothwell, R. (1992). Successful industrial innovation: Critical factors for the 1990s. *R&D Management*, 22(3), 221–240.

Rowe, G. and Frewer, L.J. (2000). Public participation methods: A framework for evaluation. *Science, Technology, and Human Values*, 25(1), 3–29.

Rowe, G. and Frewer, L.J. (2005). A typology of public engagement mechanisms. *Science, Technology, and Human Values*, 30(2), 251–290.

Seyfang, G. and Haxeltine, A. (2012). Growing grassroots innovations: Exploring the role of community-based initiatives in governing sustainable energy transitions. *Environment and Planning C: Government and Policy*, 30(3), 381–400.

Seyfang, G. and Smith, A. (2007). Grassroots innovations for sustainable development: Towards a new research and policy agenda. *Environmental Politics*, 16(4), 585

Shelley-Egan, C., Wright, D., Benčin, R., Šumič Riha, J., Strle, G., Ovadia, D., Pastor Cañedo, A., Angeli, C., Sotiriou, M. (2014). Report (handbook) of participatory processes. SATORI Deliverable D2.1 [Online] Available at: https://satoriproject.eu/media/D2.1_Report-handbook-of-participatory-processes_FINAL1.pdf [Accessed 12 April 2022].

Shelley-Egan, C., Brey, P., Rodrigues, R., Douglas, D., Gurzawska, A., Bitsch, L., Wright, D., Wadhwa, K. (2016). Ethical assessment of research and innovation: A comparative analysis of practices and institutions in the EU and selected other countries. SATORI Deliverable D1.1 [Online]. Available at: https://satoriproject.eu/media/D1.1_Ethical-assessment-of-RI_a-comparative-analysis-1.pdf [Accessed 12 April 2022].

Shrader-Frechette, K. (1995). Technology assessment. In *Encyclopedia of Bioethics*, Volume 5, 2nd revised edition, Reich, W. (ed.). Macmillan, New York.

Slocum, N. (2003). *Participatory Methods Toolkit: A Practitioner's Guide*. United Nations University, Maastricht.

Smith, A. and Seyfang, G. (2013). Constructing grassroots innovations for sustainability. *Global Environmental Change*, 23(5), 827–829.

Smith, A., Fressoli, M., Thomas, H. (2014). Grassroots innovation movements: Challenges and contributions. *Journal of Cleaner Production*, 63, 114–124.

Spina, A. (2017). A regulatory Mariage de Figaro: Risk regulation, data protection, and data ethics. *European Journal of Risk Regulation*, 8(1), 88–94.

Stahl, B.C. and Wakunuma, K. (2015). Guidelines handbook. CONSIDER Deliverable D4.1 [Online] Available at: https://docs.google.com/viewer?a=v&pid=sites&srcid=ZGVmYXVsdGRvbWFpbnxjb25zaWRlcndlYnNpdGV8Z3g6N2Q1OTUyNDRkYjJiODA3Mw [Accessed 12 April 2022].

Stanley, I., Glennie, A., Gabriel, M. (2018). How inclusive is innovation policy? Insights from an international comparison. Working paper, Nesta, London.

Steiner, J. (2012). *The Foundations of Deliberative Democracy: Empirical Research and Normative Implications*. Cambridge University Press, Cambridge.

Stilgoe, J. (2013). Why responsible innovation? In *Responsible Innovation: Managing the Responsible Emergence of Science and Innovation in Society*, Owen, R., Heintz, M., Bessant, J.R. (eds). John Wiley & Sons, Chichester.

Stilgoe, J., Owen, R., Macnaghten, P. (2013). Developing a framework for responsible innovation. *Research Policy*, 42(9), 1568–1580.

Stirling, A. (2005). "Opening up" and "closing down". Power, participation and pluralism in the social appraisal of technology. *Science, Technology and Human Values*, 33(2), 262–294.

Strand, R., Spaapen, J., Bauer, M.W., Hogan, E., Revuelta, G., Stagl, S., Paula, L., Guimarães Pereira, A. (2015). Indicators for promoting and monitoring responsible research and innovation. Report from the Expert Group on Policy Indicators for Responsible Research and Innovation. Publications Office of the European Union, Luxembourg.

Sykes, K. and Macnaghten, P. (2013). Responsible innovation – Opening up dialogue and debate. In *Responsible Innovation: Managing the Responsible Emergence of Science and Innovation in Society*, Owen, R., Heintz, M., Bessant, J.R. (eds). John Wiley & Sons, Chichester.

Taebi, B., Correljé, A., Cuppen, E., Dignum, M., Pesch, U. (2014). Responsible innovation as an endorsement of public values: The need for interdisciplinary research. *Journal of Responsible Innovation*, 1(1), 118–124.

Thapa, R.K., Iakovleva, T., Foss, L. (2019). Responsible research and innovation: A systematic review of the literature and its applications to regional studies. *European Planning Studies: Expanding the Field of Responsible Research and Innovation – From Responsible Research to Responsible Innovation*, 27(12), 2470–2490.

Tracey, P. and Stott, N. (2016). Social innovation: A window on alternative ways of organizing and innovating. *Innovation*, 19(1), 51–60.

United Nations Economic Commission for Europe (1998). Convention on Access to Information, Public Participation in Decision-Making and Access to Justice in Environmental Matters. Aarhus [Online]. Available at: https://www.unece.org/fileadmin/DAM/env/pp/documents/cep43e.pdf [Accessed 31 January 2022].

University of Montreal (2017). The Montreal declaration for a responsible development of artificial intelligence [Online]. Available at: https://www.montrealdeclaration-responsibleai.com [Accessed 31 January 2022].

Van Gorp, A. and Van der Molen, S. (2011). Parallel, embedded or just part of the team: Ethicists cooperating within a European security research project. *Science and Engineering Ethics*, 17(1), 31–43.

Van den Hoven, J., Doorn, N., Swierstra, T., Koops, B., Romijn, H. (eds) (2014). *Responsible Innovation 1: Innovative Solutions for Global Issues*. Springer, Dordrecht.

Van Lente, H., Swierstra, T., Joly, P.-B. (2017). Responsible innovation as a critique of technology assessment. *Journal of Responsible Innovation*, 4(2), 254–261.

Van de Poel, I. (2008). How should we do nanoethics? A network approach for discerning ethical issues in nanotechnology. *NanoEthics*, 2(1), 25–38.

Van de Poel, I. and Sand, M. (2021). Varieties of responsibility: Two problems of responsible innovation. *Synthese*, 198(1), 4769–4787.

Von Schomberg, R. (2007). From the ethics of technology towards an ethics of knowledge policy and knowledge assessment. Working document from the European Commission Services. Office for Official Publications of the European Communities, Luxembourg.

Von Schomberg, R. (2012a). The precautionary principle: Its use within hard and soft law. *European Journal of Risk Regulation*, 3(2), 147–156.

Von Schomberg, R. (2012b). Prospects for technology assessment in a framework of responsible research and innovation. In *Technikfolgen Abschätzen Lehren*, Dusseldorp, M. and Beecroft, R. (eds). VS Verlag für Sozialwissenschaften, Wiesbaden.

Von Schomberg, R. (2013). A vision of responsible research and innovation. In *Responsible Innovation: Managing the Responsible Emergence of Science and Innovation in Society*, Owen, R., Heintz, M., Bessant, J.R. (eds). John Wiley & Sons, Chichester.

Weyrauch, T. and Herstatt, C. (2016). What is frugal innovation? Three defining criteria. *Journal of Frugal Innovation*, 2(1), 1–17.

Wilsdon, J. and Willis, R. (2004). *See-through Science: Why Public Engagement Needs to Move Upstream*. Demos, London.

Young, I.M. (1990). *Justice and the Politics of Difference*. Princeton University Press, New Jersey.

Zwart, S.D., Van de Poel, I., Van Mil, H., Brumsen, M. (2006). A network approach for distinguishing ethical issues in research and development. *Science and Engineering Ethics*, 12(4), 663–684.

Index

A, B, C

D, E

F, G, I

J, L, M

N, O, P

R, S

T, U, V

Other titles from

iSTE

in

Interdisciplinarity, Science and Humanities

2022

ROBERT Sandrine
Resilience: Persistence and Change in Landscape Forms

2021

GUCHET Xavier
Care in Technology (Innovation and Responsibility Set – Volume 7)

2020

ATCHOUA Julien, BOGUL Jean-Jacques, DIALLO Saikou
Digital Technologies and African Societies: Challenges and Opportunities (Africa Digital Age Set – Volume 1)

CHERICI Céline
From Clouds to the Brain: The Movement of Electricity in Medical Science

GUILLAUME Astrid, KURTS-WÖSTE Lia
Making Sense, Making Science

SCHEPS Ruth, MAUREL Marie-Christine
Mathematics in the Visual Arts

PAVIE Xavier
Critical Philosophy of Innovation and the Innovator (Innovation and Responsibility Set – Volume 6)

2019

ANCORI Bernard
The Carousel of Time: Theory of Knowledge and Acceleration of Time

BOULEAU Gabrielle
Politicization of Ecological Issues: From Environmental Forms to Environmental Motives

GAYON Jean†, interviewed by PETIT Victor
Knowledge of Life Today: Conversations on Biology

KHATCHATOUROV Armen, with the collaboration of CHARDEL Pierre-Antoine, FEENBERG Andrew, PÉRIÈS Gabriel
Digital Identities in Tension: Between Autonomy and Control
(Innovation and Responsibility Set – Volume 5)

LENOIR Virgil Cristian
Ethically Structured Processes
(Innovation and Responsibility Set – Volume 4)

NIKOLOVA Blagovesta
The RRI Challenge: Responsibilization in a State of Tension with Market Regulation
(Innovation and Responsibility Set – Volume 3)

ZEMBRI-MARY Geneviève
Project Risks: Actions Around Uncertainty in Urban Planning and Infrastructure Development

2018

LOISY Catherine, COULET Jean-Claude
Competence and Program-based Approach in Training: Tools for Developing Responsible Activities
(Innovation and Responsibility Set – Volume 1)

PANSERA Mario, OWEN Richard
Innovation and Development: The Politics at the Bottom of the Pyramid
(Innovation and Responsibility Set – Volume 2)

2017

MAESSCHALCK Marc
Reflexive Governance for Research and Innovative Knowledge
(Responsible Research and Innovation Set – Volume 6)

PELLÉ Sophie
Business, Innovation and Responsibility
(Responsible Research and Innovation Set – Volume 7)

2016

GIANNI Robert
Responsibility and Freedom: The Ethical Realm of RRI
(Responsible Research and Innovation Set – Volume 2)

GRUNWALD Armin
The Hermeneutic Side of Responsible Research and Innovation
(Responsible Research and Innovation Set – Volume 5)

LENOIR Virgil Cristian
Ethical Efficiency: Responsibility and Contingency
(Responsible Research and Innovation Set – Volume 1)

PELLÉ Sophie, REBER Bernard
From Ethical Review to Responsible Research and Innovation
(Responsible Research and Innovation Set – Volume 3)

REBER Bernard
Precautionary Principle, Pluralism and Deliberation: Science and Ethics
(Responsible Research and Innovation Set – Volume 4)

Printed and bound by CPI Group (UK) Ltd, Croydon, CR0 4YY
16/09/2026
14963186-0001